TECHNOLOGY IN TRANSITION:
The 'Soo' Ship Canal, 1889-1985

Robert W. Passfield

Studies in Archaeology, Architecture and History

National Historic Parks and Sites
Canadian Parks Service
Environment Canada

Available in Canada through authorized bookstore agents and other bookstores, or by mail from the Canadian Government Publishing Centre, Supply and Services Canada, Hull, Quebec, Canada K1A 0S9.

Published under the authority
of the Minister of the Environment,
Ottawa, 1989.

Editing: Sheila Ascroft
Cover design: Rod Won

Parks publishes the results of its research in archaeology, architecture, and history. A list of publications is available from Research Publications, Environment Canada, Canadian Parks Service, 1600 Liverpool Court, Ottawa, Ontario K1A 0H3.

Canadian Cataloguing in Publication Data

Passfield, Robert W. (Robert Walter), 1942-

Technology in Transition: the 'Soo' Ship Canal, 1889-1985

(Studies in archaeology, architecture and history,
ISSN 8021-1027)
Includes abstract in French.
Includes bibliographical references and index.
ISBN 0-660-12936-1
DSS cat. no. R61-2/9-39E

1. Sault Sainte Marie Canal (Ont.) — Design and construction. 2. Canals — Ontario — Sault Ste. Marie — Design and construction. I. Canada. National Historic Parks and Sites. II. Title. III. Title: The 'Soo' Ship Canal, 1889-1985. IV. Series.

TC627.S2P37 1988   C88-097039-1   627'.137'09713132

CONTENTS

Submitted for publication in 1986, by Robert W. Passfield, Historical Research Division, National Historic Parks and Sites, Canadian Parks Service, Environment Canada, Ottawa.

SOMMAIRE

Cette étude découle d'un projet de recherche qui avait pour but d'identifier et de replacer dans un contexte historique les techniques utilisées dans la construction du canal de Sault-Sainte-Marie, Ontario, Canada, entre 1889 et 1985.

L'étude comporte trois parties qui décrivent différents aspects du canal: la construction du canal et de l'écluse (1889-1894); l'électrification et l'opération de l'écluse (1893-1985); ainsi que la conception et l'opération du barrage tournant de secours (1895-1985) qui protégeait le canal et l'écluse. Chacune des parties interprète ce qui a été construit ou mis en place au canal de Sault-Sainte-Marie dans le contexte des techniques de génie en constante évolution de l'époque. Chaque partie tente de dépeindre tous les côtés de la technologie de la puissance et de la construction reliées au projet. Chaque partie de l'étude décrit l'ensemble des facteurs externes (politiques, géographiques, économiques ainsi que techniques) qui donnèrent une qualité innovatrice à la conception, à la configuration, à l'échelle et à la nature même du canal de Sault-Saint-Marie.

L'efficacité, la durabilité et l'impact des aspects innovateurs du canal de Sault-Sainte-Marie sont analysés du temps de la construction jusqu'à maintenant. L'importance des réalisations incarnées par la construction et l'opération du canal est replacée dans le contexte contemporaine de la construction des canaux au Canada, en Amérique et en Europe occidentale.

ACKNOWLEDGEMENTS

I am indebted to Arthur C. Hudson, Electrical Engineer, National Research Council (retired), and Arnold E. Roos, historian of technology, Historical Research Branch, Canadian Parks Service, Environment Canada, for perusing earlier drafts of the manuscript.

I also wish to thank the respective staffs of the Parks Library, the Historical Research and Records Unit of the National Historic Parks and Sites, the Records Section, Ontario Region Office, as well as the Sault Ste. Marie Ship Canal Office, Canadian Parks Service, for assistance in securing research materials. I am equally indebted to the staff of the Canada Institute for Scientific and Technical Information (CISTI) Library, National Research Council.

PREFACE

The Sault Ste. Marie Ship Canal was constructed by the Canadian government in 1889-1895 to complete the last link in an all-Canadian ship navigation stretching 1400 miles from the ocean port of Montreal to the head of Lake Superior. Although only a mile long, with but a single 900 foot by 60 foot lock, the 'Soo' Ship Canal* was nonetheless a highly innovative undertaking that mirrored its age — an age of technology in transition.

By the 1890s North America was entering a new age of unfettered expansion marked in all fields of engineering by the construction of colossal structures that dwarfed earlier achievements in scale and significance. This development was made possible by the introduction of new technologies and novel adaptations of existing power and building technologies, as wood, iron, and stone masonry began to give way to structural steel and concrete, and electrical power emerged as a viable alternative to steam and hydraulic power.

All of these transitions were mirrored in the design evolution, construction, and operation of the Sault Ste. Marie Ship Canal which constituted a veritable 'touchstone' of the age. In the three major components of the 'Soo' Ship Canal — the canal with its electrically powered lock, the hydro-electric canal powerhouse, and the emergency swing bridge dam protecting the canal and lock — new technologies were introduced and older technologies employed in novel adaptations.

On the Sault Ste. Marie Ship Canal project, structural steel was widely used in place of the more traditional wood and iron. Concrete replaced the traditional clay puddle cut-off walls. It was also used structurally for the first time in Canadian canal construction to supplement the conventional timber crib and stone masonry retaining wall, and to construct novel lock chamber walls of rubble concrete faced with cut stone.

* In this book, 'Soo' Canal and 'Soo' Ship Canal are synonymous with the Sault Ste. Marie Ship Canal in Ontario, Canada. The American canal at Sault Ste. Marie, Michigan, is denoted by its proper name of St. Mary's Falls Ship Canal.

The Great Lakes – St. Lawrence

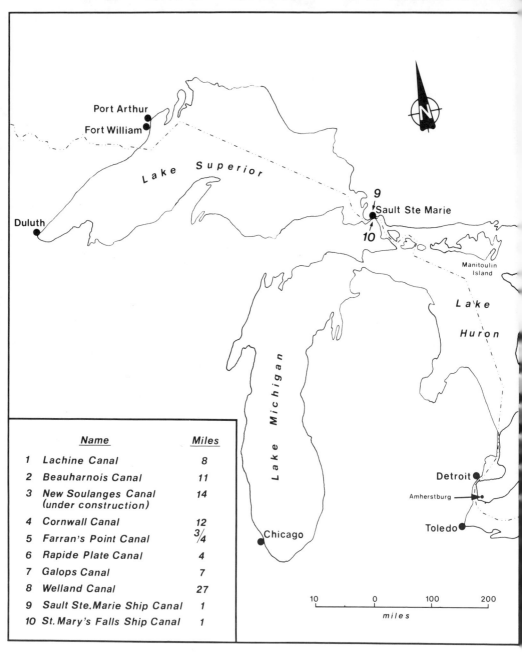

Name	Miles
1 Lachine Canal	8
2 Beauharnois Canal	11
3 New Soulanges Canal (under construction)	14
4 Cornwall Canal	12
5 Farran's Point Canal	3/4
6 Rapide Plate Canal	4
7 Galops Canal	7
8 Welland Canal	27
9 Sault Ste.Marie Ship Canal	1
10 St. Mary's Falls Ship Canal	1

Port Arthur
Fort William
Lake Superior
Duluth
Sault Ste Marie
Manitoulin Island
Lake Huron
Lake Michigan
Detroit
Amherstburg
Toledo
Chicago

10   0   100   200
miles

Transportation System, 1895

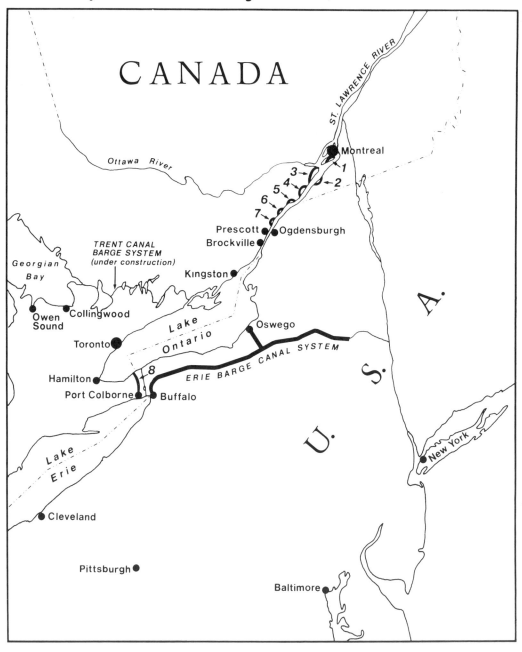

CANADA

ST. LAWRENCE RIVER

Ottawa River

Montreal

3
1
4
2
5
6
7

Prescott
Brockville
Ogdensburgh

Kingston

TRENT CANAL
BARGE SYSTEM
(under construction)

Georgian
Bay

Owen
Sound
Collingwood

Lake
Ontario

Oswego

ERIE BARGE CANAL SYSTEM

Toronto

A.

Hamilton
Port Colborne
8
Buffalo

U. S.

Lake
Erie

New York

Cleveland

Pittsburgh

Baltimore

Doug Sullivan
Planning Officer

National Historic
Parks and Sites

Hydraulic power, susceptible to cold weather operating difficulties, was rejected in favour of a novel adaptation of a newly established d.c. electrical power for operating the large scale lock. Moreover, the entire canal was illuminated using d.c. lighting technology — incandescent lighting for indoors, and arc lighting outdoors. When difficulties were experienced with the lighting system, consideration was immediately given to converting the whole canal to the comparatively new polyphase a.c. system. In 1906, the arc lighting was converted to the a.c. system, employing transformers to overcome distribution problems, and the incandescent lighting system was converted in 1921.

In its scale and configuration, the 'Soo' Ship Canal lock was also a response to the mammoth steel ships that began to supersede iron vessels and the remnants of an earlier age of wooden sailing ships, in carrying cargo tonnages increasing at a rate far in excess of the most sanguine projections.

At Sault Ste. Marie, Canadian engineers developed a highly efficient canal utilizing a novel configuration of flotilla lock and unique operating system. On its opening in September 1895, the 'Soo' Ship Canal had the world's first electrically powered lock, by far the world's longest lock, and was the first North American canal to be totally illuminated with electric lighting. The highly efficient lock was for a time the world's busiest on a canal that in 1913 set a then-world record in passing 42.7 million tons of freight, far surpassing the more renown Suez Canal.

A novel emergency swing bridge dam, developed to protect large-scale ship canals, also received its major, and only test at Sault Ste. Marie. The successful operation of the emergency dam in closing off a raging torrent of water, after a potentially catastrophic accident in June 1909, proved its efficacy and influenced the design of emergency dams erected subsequently on the Panama Canal.

The ship canal was converted to polyphase a.c. power on the regional power grid in 1942. Subsequently, the canal powerhouse was stripped of all its electrical power generating equipment. Only one turbine and the pumping machinery for dewatering the lock chamber remained. The lock, however, continued to operate with its original machinery intact, and protected by the emergency swing bridge dam — the sole remnant of its type. In 1979 commercial operation ceased, and thereafter Parks Canada (now Environment Canada, Canadian Parks Service) continued the canal in operation as a recreational waterway for pleasure craft. The canal was forced to close during the summer of 1987 after a major crack opened in the south wall of the lock.

Robert W. Passfield
Ottawa
February 1988

I. CONSTRUCTING THE CANAL AND LOCK, 1889-94

Introduction

In 1889-95, the Canadian government constructed a ship canal at Sault Ste. Marie, Ontario, to complete the last link in an all-Canadian canal transport system stretching more than 1400 miles from the ocean port of Montreal through the Great Lakes to the head of Lake Superior. As initially planned, the single lock of the mile-long ship canal was to have been hydraulically operated and built on a scale in keeping with ship locks erected elsewhere on the Canadian ship canal system. Changes were made, however, which resulted in the introduction of several major innovations in Canadian canal building technology. These included an innovative use of concrete, and a new configuration of ship lock that led to the construction of by far the world's longest lock.

The highly efficient long lock had a major influence on ship canal construction, and for a time, would play a major role in establishing the Sault Ste. Marie waterway as the world's busiest ship canal system. This subsequent success rested on the various changes and innovations introduced during the construction period. This was done in response to a complex web of political, economic, and commercial factors acting on the Canadian government during a period of growing industrialization and rapid technological advances. All of these factors came to bear with dramatic results during the construction of the canal and lock chamber of the Sault Ste. Marie Ship Canal in 1889-94.

Enlarging the Canadian Canal System, 1871

When the government of Sir John A. Macdonald authorized the construction of a ship canal at Sault Ste. Marie in 1888,[1] it was but one aspect of a much broader policy of national economic development, expansion and integration. A decade earlier, the Conservatives had begun implementing what was termed the "national policy." It envisaged the settlement of the newly acquired western territories through encouraging immigration, the development of the western prairies as a major world's wheat growing and exporting region, the fostering of industrialization in the East through tariff protection for nascent Canadian industries, and the construction of a transcontinental railway to carry Canadian manufactured goods to the prairie farmers and prairie wheat to world markets.

For almost two decades prior to 1878, the rock of the Precambrian Shield along the upper Great Lakes had curtailed Canadian agricultural expansion. By means of the national policy program, the Conservatives planned to leap over that obstacle to the open prairies beyond and develop a transcontinental economy based on wheat.[2]

A Canadian transcontinental railway was the key to westward expansion and national integration, but the enlargement and extension of Canada's canal system was a no less critical component.[3] If Canadian wheat was to be competitive on world markets, cheap transportation was essential. Wheat prices and the heavy ocean freight rates from Montreal were beyond government control,[4] but inland railway freight costs could be dramatically reduced by substituting water transportation wherever possible.

Owing to cost factors, Canadian grain shippers had traditionally favoured water over rail transport.[5] Wheat shipped from Toronto to Montreal by rail cost anywhere from 50 cents to $1.50 a ton more than on the St. Lawrence canal system,[6] and this discrepancy could only increase dramatically with the greater distances involved in shipping prairie wheat. Indeed, it was doubtful that wheat grown on the Canadian prairies could be shipped all the way to Montreal by rail without the total transportation costs, rail plus ocean freight, absorbing almost the market value of the produce. The Americans, however, had found that western wheat could be shipped from Duluth and Chicago through the Great Lakes at one-tenth to one-eighth the cost of rail transport to Buffalo where it was transshipped into Erie Canal barges or rail cars for the last 300 miles of the journey to New York.[7] Hence the Conservatives planned to revive a canal enlargement program that had its origins in an earlier rivalry between the Canadian St. Lawrence transportation system based on Montreal and the American Erie Canal system centred on New York.

Prior to 1825, the St. Lawrence commercial system had dominated the trade of the Great Lakes interior, but lost that pre-eminence to New York on the opening of the Erie Canal. During the 1840s, a concerted effort had been made to re-establish the hegemony of Montreal through the construction of a series of canals by-passing the rapids of the St. Lawrence River. The St. Lawrence canals had cut transportation costs in half, enabling Canadian produce to continue to compete in British markets, but the decline in shipping costs was insufficient to overcome the cost advantages enjoyed by New York as a major importing, as well as exporting, centre in the transatlantic trade.[8]

A second effort to recapture the trade of the American agriculture frontier for the St. Lawrence commercial system had seen the construction, during the 1850s, of the Grand Trunk Railway (GTR) from Montreal through to Sarnia and beyond to Detroit. The GTR and its feeder railways had succeeded in tying the Canadian hinterland trade to the St. Lawrence, and brought the establishment of heavy industries. American traffic, however, was minimal owing to the parallel development of American railway systems linked to New York and the comparatively low ocean freight rates it continued to enjoy.[9]

As of 1871, it had been clear to the government of the new Canadian confederation that trunk railways and the existing canal system would not suffice to capture a major share of the rapidly developing trade of the American Midwest for Canadian trade channels. The only viable alternative was a uniform system of inland water transportation that would eliminate transshipments on the Welland and St. Lawrence canals, and enable the large lake freighters to pass directly from Lake Superior through to Montreal. Consequently, Sir John A. Macdonald's government had embarked on a major canal construction program. It involved enlarging the six existing St. Lawrence canals — the Lachine, the Beauharnois, Cornwall, Farran's Point, Rapide Plat, the Galops — as well as the Welland Canal by-passing Niagara Falls. A proposed new canal, the Sault Ste. Marie Canal, would complete the system. It was to be constructed at the St. Mary's Falls where the waters of the St. Mary's River, in flowing between Lake Superior and Lake Huron, narrowed to about 2300 feet in width and became less than three feet deep. There the river fell 18 feet in a distance of half a mile across a boulder strewn bedrock that impeded all navigation.

All of the canals were to be constructed on a uniform scale with a 12-foot depth of navigation, and equipped with large 270 foot by 45 foot ship canal locks.[10] In all, a total of 72 miles of canal were to be excavated on a 1400-mile-long ship navigation stretching from Montreal through to the head of Lake Superior.[11]

In 1873, work had commenced on the Lachine and Welland canal enlargements; but the fall of the Conservative government in November 1873, and the onset of a severe world-wide economic depression, had slowed the implementation of the canal enlargement program. Although committed to retrenchment, the new Liberal government had undertaken one additional work — the enlargement of a major section of the Cornwall Canal in 1876 — and made a commitment to increase the depth of draught in the new canal system to 14 feet.[12] Then, with the return of the Conservatives to power in 1878, the whole focus of the proposed canal transportation system had shifted to the Canadian prairies in keeping with the objectives of the new "national policy." Initially, the Conservatives had concentrated on getting the transcontinental railway under construction and the implementation of a protective tariff.[13] The canal enlargement program, however, was not forgotten.

As of 1880, with the economic outlook improving, the Conservatives had begun once again to push forward ship canal construction. The first Cornwall Canal contract was completed in 1882, the Lachine Canal enlargement in 1884, and the Welland Canal enlargement in 1887. In the interim, the remaining work on the Cornwall Canal was contracted out, and a survey was undertaken for the proposed Sault Ste. Marie Canal. In November 1888, contracts were let for the enlargement of the Rapide Plat and Galops canals, and for the construction of the new Sault Ste. Marie Canal, leaving only two canals untouched: the Farran's Point Canal; and the

Beauharnois Canal.[14] The government, however, was committed to completing the whole canal enlargement program by 1895.[15]

During the summer of 1888, the major concern was to determine the lockage capacity required for the new mile-long canal at Sault Ste. Marie: a single lock 270 feet by 45 feet; two parallel single locks on a similar scale; or possibly a large "flotilla" lock capable of passing three or four vessels at a single lockage.[16] Before deciding, Canadian canal engineers examined the existing American canal system at Sault Ste. Marie.

The American 'Soo' Canal Phenomenon

At Sault Ste. Marie, the state of Michigan had constructed a canal in 1853-55 to provide a cheap all-water transportation system for shipping out ore from the rich copper and iron mines of the upper peninsula.[17] The St. Mary's Falls Canal was one-mile long, with a 12-foot depth throughout. It surmounted the 18-foot fall of the St. Mary's Falls by means of two combined locks, each a nine-foot lift. The two 350 foot by 70 foot masonry locks were the first ship canal locks constructed in the United States, and were by far the largest ship locks in North America. The canal had cost the State $999 802 but proved a phenomenal success.[18]

In the first year of navigation, 14 503 tons of freight had passed through the St. Mary's Falls Canal, and within five years shipments had increased tenfold with a tonnage of 153 721 tons being recorded for the 1860 navigation season.[19] Initially copper had constituted the bulk of the freight, but as the Michigan iron mines increased production and new mines were opened south of Lake Superior, iron ore had rapidly eclipsed copper. Ore shipments had boomed. In 1855, only 1449 tons of ore were shipped via the St. Mary's Falls Canal, but this had increased to 114 401 tons in 1860, reached 830 940 tons in 1870, and 1 908 745 tons in 1880. Indeed, in the course of just over two decades the St. Mary's Falls Canal had become the lifeline of the rapidly expanding American iron and steel industry, and the Michigan State locks had been found inadequate to meet the rapidly increasing needs of commerce.[20]

As early as 1876, the United States government had undertaken the construction of a huge new lock — the Weitzel lock — parallel to the existing combined locks to dramatically increase the carrying capacity of the St. Mary's Falls Canal. During the winter of 1880-81, the canal prism was widened, from 100 to 108 feet at the waterline, and deepened to 16 feet throughout. On opening in September 1881, the so-called "Colossus of Locks" was by far the largest lock in the world. It was a flotilla lock designed to pass four vessels in a single lockage. The chamber measured 515 feet in length by 80 feet in width, narrowing to 60 feet at the gates with 17 feet of water on the sills.[21] The Weitzel lock had far exceeded the capacity of the 270 foot by 45 foot locks being constructed contemporaneously on the Lachine, Cornwall, and Welland canals on the projected 14-foot-deep Canadian ship canal navigation system. And yet, within a few years it was evident that the new St. Mary's Falls Canal — enlarged at a cost of $2 150 000[22] — would soon be inadequate to meet growing shipping demands.

Prior to 1869, all bulk freight on the upper lakes had been carried in wooden hulled sailing schooners, the largest of which were 175 feet in length with a 35-foot beam. Steamboats — both sidewheelers and the propellers — carried only passengers and package freight. The early steamers were too confined for bulk freight owing to the large arched trusses, so-called "hogging trusses," which ran along each side of the deck to stiffen shallow hulls ranging anywhere from 230 to 300 feet in length (Fig. 1). Within 20 years of the opening of the St. Mary's Falls Canal, however, new ship configurations and the introduction of iron and then steel hulls, had initiated a long trend toward ever increasing dimensions for lake freighters.

FIGURE 1. *The Lady Elgin* built at Buffalo, N.Y., for the Grand Trunk Railway, 1851. A typical passenger/package freight steamer in the mid-1850s with longitudinal "hogging" trusses strengthening the hull. (Metropolitan Toronto Reference Library, J. Ross Robertson Collection, T-16166)

In 1869, the first "steam barge" had been launched at Cleveland. The *R.J. Hackett* was given a boxy hull similar to that of the large lake schooners, but with the arched stiffening trusses of the steamboat built into the sides of the hull. To further clear the deck, the pilot-house was moved far forward and the operating machinery far aft — the classic design for subsequent generations of lake freighters (Fig. 2). The new steam barges had proved very successful in the bulk cargo trade

FIGURE 2. The lake boat, *S.S. Forest City*, has classic boxy hull with arched stiffening trusses built into its sides and an uninterrupted deck, introduced into steam freighter design by the *R.J. Hackett* in 1869. (Andrew E. Young, National Archives of Canada, PA 145465)

with vessels up to 230 feet in length being constructed; and sailing schooners, to remain competitive, had followed suit attaining over 200 feet in length.[23]

The Weitzel lock colossus had been designed to accommodate the big upper lakes freighters, but it had no sooner been built than even larger vessels appeared. In 1881 a sailing schooner, 265 feet in length with a 38-foot beam, had been launched at Toledo. The *David Dows* had a capacity of 140 000 bushels of grain, and for a time was the largest schooner in the world. It was superseded in size the following year when the first iron-hulled steam barge, the 282-foot *Onoko*, was launched at Cleveland (Fig. 3). Four years later, in 1886, the first of the steel-hulled lake freighters, the 310-foot *Spokane*, appeared at the same port.[24] In effect only two of these vessels at a time, rather than four freighters as previously, could pass through the Weitzel lock in a single lockage. Not only were freight vessels becoming increasingly large, but they were being built in far greater numbers to accommodate the bulk cargo trade on the upper lakes.

FIGURE 3. Painting of the steamer *Onoko*, the first iron-hulled lake boat launched in 1882. (*Ships of the Great Lakes; a Pictorial History*, courtesy of Karl Kuttruff)

At Sault Ste. Marie, the bulk shipping tonnage consisted of ore, coal, grain and lumber. Eastbound vessels carried ore, grain and lumber. General merchandise and coal constituted the return freight which totalled approximately half the tonnage of eastbound traffic[25] Iron ore, however, predominated. The 1885 navigation season saw 3 256 628 tons of freight pass through the St. Mary's Falls Canal, with iron ore shipments comprising 2 466 372 tons.[26] In that year shipping reached almost half the annual tonnage on the Suez Canal, the world's premier ship canal.[27] Moreover, freight tonnage on the St. Mary's Falls Canal had been increasing at the amazing rate of 107 000 tons a year.[28]

Shipping projections showed that the capacity of the canal would be exceeded within a decade, and to meet that crisis the United States government had undertaken to establish a 20-foot-deep ship navigation from Duluth on Lake Superior through to Buffalo on Lake Erie. This project involved the deepening of both the Detroit-St. Clair and St. Mary's rivers, including the opening of a new American deep-water channel through Hay Lake below the St. Mary's Falls Canal, and the deepening of the canal in conjunction with the construction of a larger lock. The mammoth new lock, on which construction had commenced in 1887 on the site of the old Michigan State combined locks, would supersede the Weitzel lock as by far and away the largest lock in the world.

The new Poe lock, scheduled for completion in 1897 at an estimated cost of $4 738 865, measured 800 feet by 100 feet with 21 feet of water on the sills.[29] It was also a flotilla lock designed to pass four of the then-largest upper lakes steam barges in a single lockage.[30] And such a capacity was by no means excessive.

Canadian engineers noted that as many as 84 vessels passed through the American canal in a 24-hour day, and over the whole month of June 1887 an average of 56 daily vessel passages was maintained.

In assessing the American experience at Sault Ste. Marie, it was difficult for Canadian canal engineers to determine the scale of canal best suited to meet future Canadian needs. Questions abounded. Would ships continue to increase in size? Would the volume of traffic generated by the settlement and development of the resources of the Canadian Northwest approximate the American experience? Should the new Canadian canal at Sault Ste. Marie seek to match the projected 20-foot American ship navigation or maintain the existing scale of the on-going Canadian ship canal enlargement program? Who could predict what was required when, as was noted in April 1888, traffic on the American St. Mary's Falls Canal:

> far exceeds what was a few years ago looked upon as the inflated views of impracticable persons dealing in such forecasts as were unwarranted by time or circumstances.[31]

The Canadian 'Soo' Canal Undertaking (1887-88)

As of 1880, when the Conservative government began to push forward the canal enlargement program, the potential for wheat production on the Canadian prairies had appeared almost limitless.[32] It was widely estimated that at least 150 million acres were suitable for cultivation, and settlement was rapidly spreading onto the prairies. Some 132 918 acres of land had been taken up in 1876, increasing to over a million acres in 1879, and attaining 2 699 144 acres by 1882 in the midst of a veritable land boom. Predictions were being made that within a generation wheat production on the Canadian prairies would exceed the total of the new American western states.[33] There the spread of settlement into Illinois, Wisconsin, Minnesota, Iowa, Nebraska, and Kansas, had helped raise the annual American wheat production from 176 to 459 million bushels over the 20 years prior to 1880.[34]

While still sparcely settled, Manitoba alone in 1878 had produced over a million bushels of hard northern wheat, and began to ship grain, in ever increasing volumes, to the south over the American railway system and beyond over the lakes to Toronto. During the summer of 1883, the progress of construction of the new Canadian Pacific Railway (CPR) had opened up a line from Port Arthur on Lake Superior through to Winnipeg and beyond almost to Calgary. This had marked the beginning of bulk shipments of western wheat over the Canadian transcontinental railway for export via Port Arthur and the Great Lakes-St. Lawrence system.[35]

To move immigrants and package freight west, the CPR as of 1884 placed a fleet of exceptionally large, first-class steamers in regular service between its rail-

FIGURE 4. The *Athabasca*, one of three identical passenger steamers placed in service on the upper lakes by the Canadian Pacific Steamship Line in 1884. (National Archives of Canada, C-4861)

way terminal at Owen Sound on Georgian Bay and Port Arthur and nearby Fort William on Lake Superior, via the American St. Mary's Falls Ship Canal. Three new steamers, each 263 feet in length with a 38-foot beam, were employed: the *Algoma, Alberta* and *Athabaska* (Fig. 4).[36]

While engaged in railway construction, the CPR in 1883-85 had also established a system of grain elevators to handle the anticipated outpouring of western wheat. An elevator of 650 000 bushels capacity was constructed at Port Arthur, another elevator of 1 350 000 bushels capacity in Fort William, and two elevators with a total capacity of 1 200 000 bushels at Montreal. With the St. Lawrence canal enlargement program as yet incomplete, another elevator was acquired at Brockville just upstream of the existing St. Lawrence canal system for transshipping the grain into barges for conveyance to Montreal. Moreover, a second system for moving western wheat, combining water and rail transport, had been established through constructing an elevator of 250 000-bushels capacity at Owen Sound. It would enable lake freighters departing from Port Arthur-Fort William to turn off the main shipping route, after passing through the proposed Sault Ste. Marie Canal, and enter Georgian Bay to transship their cargoes into the existing CPR Railway network for transport from Owen Sound to Montreal.

As planning proceeded for the Canadian Sault Ste. Marie Canal, western grain shipments were growing rapidly. In 1885, some 7 842 343 bushels of grain, mostly wheat, was carried east by the CPR. This had increased to 10 960 582 bushels in 1886, and 15 013 957 the next year following an exceptionally good harvest.[37] But despite such promising beginnings, western development was not proceeding as anticipated. In 1883, the western land boom had collapsed, initiating a period of declining wheat prices and a serious depression in the West.[38] With the completion of the CPR system from coast to coast in 1886, immigrants had been expected to flood onto the prairies. The tide of European emigration, however, was continuing to flow into the American West ignoring the Canadian prairies.[39]

The disappointingly slow growth of the Canadian population was insufficient to people the Canadian West in the foreseeable future, and threatened to undermine one of the key elements of the national policy program.[40] What was worse, settlers were experiencing difficulties in adapting to prairie farming. Large numbers were abandoning their homesteads, joining an already heavy Canadian emigration into the United States.[41]

Despite such setbacks, faith persisted in the vast potential of the western lands.[42] The Conservative government and the canal engineers designing the Sault Ste. Marie Canal were by no means pessimistic. They were convinced that wheat shipments from the Canadian West would continue to expand at an ever increasing rate, adding enormously to the trade of the Great Lakes-St. Lawrence system. Hence it was essential that the Sault Ste. Marie Canal be constructed on a scale sufficient to meet all future demands.[43]

After assessing the American St. Mary's Falls Canal phenomenon, the potential of the Canadian grain trade, and the on-going Canadian canal enlargement program, a decision was reached as to the most appropriate scale for the new canal at Sault Ste. Marie. On the upper lakes, this was a navigation capable of passing vessels of a 16-foot draught. Hence, plans were prepared for an 18-foot deep canal, having a single flotilla lock 600 foot by 85 feet with 16 feet three inches of water on the sills at extreme low water. The lock would be capable of passing two of the largest freighters is the Canadian lake fleet at one lockage.[44] These were the so-called "canallers" or "Welland Canallers" that had come into existence following the completion of the Welland Canal enlargement in 1887 to carry bulk freight, as well as passengers and package freight, through from the upper lakes to Montreal in anticipation of the completion of the enlarged Canadian canal system (Fig. 5)

In the existing grain trade, two different scales of vessels were used depending on their destinations. The upper lakes boats — the large sailing schooners over 200 feet by 38 feet and steam barges 320 feet by 43 feet — were employed for the carriage of wheat and flour from Duluth and Chicago to Buffalo at the head of the Erie Canal barge system. Grain and flour proceeding beyond Buffalo was carried

FIGURE 5. The *Algonquin*, a typical "Welland Canaller" lake boat. (National Archives of Canada, PA-151152)

in the "Welland Canallers" — steam vessels 255 feet long with a 42-foot beam. They were loaded at Duluth, Port Arthur-Fort William, or Chicago with 2300 tons of grain, or flour, on a 15-foot draught to pass through the American St. Mary's Falls Canal and St. Clair-Detroit rivers at Port Colborne at the head of the Welland Canal. There, part of the cargo was transshipped into railway cars and the lake boats, lightened to about 1825 tons on a 14-foot draught, passed on through the enlarged Welland Canal. The American boats then proceeded to Oswego to transship into the Erie Barge Canal or to Ogdensburg on the New York railway system. Canadian grain traffic went to Kingston, Brockville or Prescott to transship into barges going down through the St. Lawrence canals system to Montreal.[45]

The proposed Canadian 'Soo' lock was larger than the existing 515 foot by 80 foot Weitzel lock on the St. Mary's Falls Canal, but by no means as large as the 800 foot by 100 foot Poe lock. The mammoth new American lock was being constructed primarily for the iron ore, coal, and grain trade carried in the large "upper lakes boats"; whereas the size of Canadian vessels was limited by the scale of the enlarged Welland and St. Lawrence canals. The draught, however, was not so limited for the upper lakes owing to the lightening system in existence at Port Colborne. Hence the carrying capacity of the proposed Canadian 'Soo' canal was to be maximized by increasing its depth two feet beyond the 14-foot draught standard set for the enlarged Welland-St. Lawrence canals system.[46]

Although the Americans were engaged in deepening the St. Mary's and St. Clair-Detroit rivers to take vessels of 20-foot draught, and were opening a new channel of like depth through Hay Lake in American waters below the St. Mary's

Falls Ship Canal, the Canadian government had several reasons for choosing a 16-foot standard for the new Sault Ste. Marie Canal.[47] None of the Canadian harbours on the lakes above the Welland Canal was capable of taking vessels of a 20-foot draught, and the nature of the Canadian channel in the lower St. Mary's River rendered the cost of deepening beyond 17 feet prohibitive.[48]

The 16-foot draught of the proposed 'Soo' Canal would enable the existing Welland Canal boats to carry anywhere from 3000 to 3500 tons of grain from Port Arthur-Fort William to Port Colborne. There the vessels could be lightened before continuing on through the Welland Canal.[49]

As of the summer of 1888, the government had decided on what appeared to be the optimum lock size and depth of navigation for the new Sault Ste. Marie Canal, but a new question arose: Ought the Canal to be constructed? When the project was submitted to Parliament, a rancorous debate ensued. Opposition critics charged that the 'Soo' canal would cost three times the $1 million preliminary estimate, and was a waste of money. Canadians had already been granted the use of the St. Mary's Falls Ship Canal on equal terms with American citizens, and the new American Poe lock and 20-foot canal navigation system under construction would suffice for Canadian traffic needs at Sault Ste. Marie for years to come. Indeed, it was charged that the government:

> *are undertaking a work not demanded by the commercial necessities of this country; they are undertaking a work when the bill is already filled by the American Government.... There is no more necessity for the construction of this canal than there is for a jug to have two handles....*[50]

Despite such criticisms, the government remained adamant that the national interest demanded a canal. Although Canadian shipping accounted for no more than three percent of the freight tonnage on the St. Mary's Falls Canal, a commitment had been made to the construction of an all-Canadian transcontinental transportation system. If the objectives of the national policy were to be attained, a key link in the east-west system could not be left subject to foreign control.[51] Two incidents had already shown the danger in this situation.

To encourage grain ships passing through the Welland Canal to continue into the St. Lawrence transportation system, the Conservative government had instituted a shipping rebate system. All freight passing through the Welland Canal paid a toll but vessels proceeding to Canadian ports, rather than to Oswego or Ogdensburg, received a substantial rebate. This had infuriated American shipping interests and, as of the summer of 1888, the American government was threatening to impose retaliatory tolls on Canadian shipping at Sault Ste. Marie.[52]

With the projected growth of the Canadian grain export trade, the American threat was by no means a minor concern. It also revived memories of an earlier episode when the Americans, through control over the canal at Sault Ste. Marie, had acutely embarrassed the Canadian government.

In the spring of 1870, troops had been dispatched to the Northwest to restore order following the Riel rebellion in Manitoba. The Red River expedition was to proceed on the steamers *Algoma*, an earlier namesake of the CPR steamer, and *Chicora* to Fort William at the head of Lake Superior. The *Algoma* had passed through the St. Mary's Falls Ship Canal, without incident, but by the time the *Chicora* arrived the American government had acted to deny passage to any vessel carrying troops or war matériel (Fig. 6). Much to the consternation of Canadians, the expedition was delayed a week while the troops and their equipment were unloaded and moved across the old mile-long fur brigade portage on the Canadian side of the St. Mary's River.

FIGURE 6. The Chicora sidewheel steamer, used to transport troops westward during the Red River expedition of 1870. (National Archives of Canada, C-48869)

Almost two decades later, memories of the "Chicora incident" still rankled and were evoked in defence of the government's plans for a Canadian canal at Sault Ste. Marie. As Sir Charles Tupper, the Minister of Finance, exclaimed: "the time has come when we should not be in the humiliating position in which we found ourselves before...." But independence had its price. As planning proceeded, the canal estimate was increased to a more realistic $2.8 million.[53]

Initial Lock Design and Construction (1888-90)

The 600 foot by 85 foot flotilla lock was to be constructed on St. Mary's Island on the Canadian side of the river adjacent to the St. Mary's Falls. The canal ran the full length of the island, about two-thirds of a mile, and the canal prism — 150 feet wide and 18 feet deep at extreme low water — was to be excavated for the most part through a red sandstone bedrock of a seemingly porous nature.[54] The lock was positioned at the lower end of the canal, and given an exceptionally high lift of 18 feet — a lift comparable to the American Weitzel lock and the new Poe lock under construction at the 'Soo', but far greater than any lift lock on the Welland-St. Lawrence ship canal system (Fig. 7).[55]

The proposed lock was for the most part of a conventional design with stone masonry walls faced with cut stone and mitre gates of wood construction. But it differed in several significant features from what had been constructed previously on Canadian ship canals. These design innovations arose from a concern over the monumental size of the new Canadian lock on a canal opening into Lake Superior: the largest freshwater lake in the world, some 350 miles long and 100 miles wide, comprising a surface area of over 30 000 square miles.[56] Any lock gate failure would release a wall of water 18 feet deep by 150 feet wide contracting and flowing in a torrent down through the lock chamber — a potential disaster to be avoided at all costs. The new lock features, not surprisingly, reflected what the Americans had incorporated earlier into the design of the Weitzel lock colossus in response to similar concerns.

The chamber of the proposed flotilla lock differed from any previous Canadian lock in that the gates did not extend the full width of the lock. They were off-set at opposite corners of the lockchamber. One side wall curved inwards at each end, narrowing the lock from 85 feet in width down to 60 feet wide at the gate openings.[57] This configuration of lock chamber had been adopted by the Americans as it reduced the gate openings, and hence the danger of the huge wooden mitre gates failing under stress.[58] The upper gates of the Canadian lock were also to be positioned upon a breastwork, as in the Weitzel lock, marking the re-introduction of this gate arrangement into Canadian lock construction.

The shorter upper gates, mounted on a breastwork, were a standard feature of canal locks constructed almost everywhere else in the world, and had been employed in early Canadian lock construction. During the construction of the first Welland Canal in 1824-29, however, a new configuration had been introduced, and

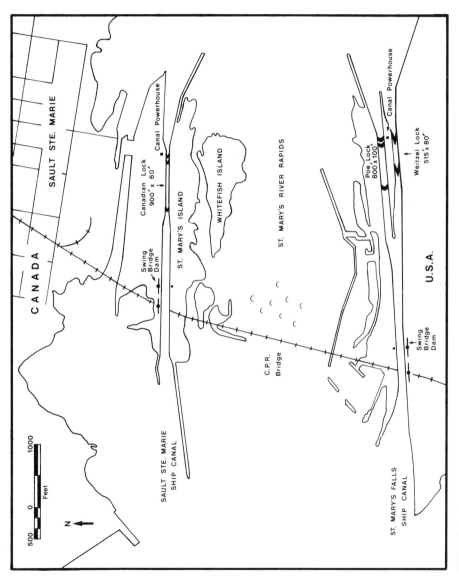

FIGURE 7. Map of the Canadian and American ship canals at Sault Ste. Marie - 1896.

used in building the St. Lawrence canals of the 1840s and during the canal enlargements of the 1870s and 1880s. In the Welland Canal arrangement, the upper and
lower gate sills were on the same level, and the breastwork was constructed in advance of the upper gates to step up to the next canal reach. This arrangement dispensed with culvert sluices in the lock wall in favour of the far simpler gate sluices.

In the standard lock configuration, with the shorter upper gates mounted up
on a breastwork, sluices could not be installed in the upper gates. Water shooting
through the upper gates into the lock chamber would flood the cargoes of ascending vessels. Therefore culvert sluices were invariably constructed to carry water
down and around the upper gates through the lock wall masonry and into the lock
chamber at its floor level. With the Welland Canal configuration, the upper gates
extended down to the level of the lock chamber floor. Water could be let into the
lock through sluices in the gates without any danger of flooding ascending boats
and barges.[59]

The major drawback of the Welland Canal lock configuration was that it greatly increased the water pressure acting against the upper gates. Consequently it was
decided to adopt the more conventional lock configuration at the upper gates of the
Sault Ste. Marie lock. This decision in turn necessitated the construction of some
type of sluice culvert system in the lock masonry for filling the lock chamber.

For both filling and emptying the lock chamber, a floor culvert system was
adopted.[60] In the standard sluice system with culverts passing around the upper
gates, as well as in the Welland Canal system with sluices directly in the gates, the
turbulence on the water entering the head of a lock was a serious problem. The
gushing water tended to drive the vessel in the lock chamber back against the lower
gates, occasionally breaking lines and causing damage to both boat and lower gates.
In the floor culvert sluice system, the water was let into the lock chamber through
apertures along the top of the culverts, thereby lifting the boat along its keel rather
than shoving it backwards.[61] This way the water could be let into the lock chamber at a much faster rate — a serious concern given the exceptionally heavy traffic expected on the Sault Ste. Marie Canal.

To further speed up lockages, a hydraulic system was planned for operating
the sluice culverts and lock gates. Both the floor culvert sluice concept and the
hydraulic system of lock operation were innovations that had been introduced earlier on the American 'Soo' canal.

To operate the Weitzel lock, a powerhouse had been constructed adjacent to
the lower gates with a 36-inch-diameter penstock to carry water from the upper
reach of the canal to two 30-inch, 50-horsepower turbines. The turbines, operated
by the 18-foot fall, were belted to two force pumps pumping water into an accumulator where it was held at about 120 p.s.i., maintaining pressure in the piping
system until needed to work the hydraulic gate and valve cylinders.[62] To date, no

Canadian lock had been equipped with hydraulic gates and sluices. But with such a system of operation, and the novel floor culvert sluice system, the American lock gates could be closed in 1.5 minutes and the huge lock chamber filled in only 12 minutes, or emptied in 7.5 minutes.[63] Such a system appeared admirably suited to the Canadian need for a highly efficient ship canal lock at Sault Ste. Marie.

In November 1888 the contract for the canal, lock and powerhouse was awarded to Hugh Ryan and Company of Toronto, a railway construction company that had built over 200 miles of the recently completed Canadian Pacific Railway west from Fort William. With the canal scheduled for completion as of 10 May 1892 — just over three years hence — no time was to be lost.[64] Work commenced in May 1889, when Ryan's men began excavating a layer of sandy loam mixed with boulders that overlay the bedrock of red Potsdam sandstone on St. Mary's Island.[65]

In anticipation of severe water seepage problems, and extensive rock excavation work, Ryan immediately established a power plant. A dam was erected across a narrow channel between the Whitefish and St. Mary's islands, and a penstock constructed to a temporary powerhouse. The head of water so obtained was used to drive five 52-inch turbines capable of generating a total of 1000 horsepower. One turbine operated a cable-power system by which a large pump engine was run for dewatering the lockpit and canal prism excavations. The cable system also ran a stone crusher about 1300 feet away. A second turbine drove line shafting in a machine shop erected on site. The remaining three turbines drove air compressors for operating rock drills and hoisting derricks.[66]

At Sault Ste. Marie, air compressors were employed for the first time on a Canadian public works project, but had been used previously in mining and for powering machinery in railway shops.[67] For many applications, compressed air had proved far superior to hydraulic power. It could be transmitted over long distances of widely differing elevations, with little perceptible pressure loss, was not susceptible to freezing in winter, and after actuating machinery could be easily dissipated back into the air.[68] In the late 19th century mechanical air compressors, utilizing a piston to compress air in a cylinder, were generally steam driven but turbine powered compressors were by no means novel.[69] During the winter of 1888-89, however, fluctuations in the river level and ice problems led Ryan to install steam power to supplement the compressed air system.[70] A dynamite factory was also established on site to facilitate the rock excavation work.[71] As this proceeded, a change was made in the lock masonry.

According to specifications, the cut stone facing of the lock walls was to be of limestone blocks laid in courses 18 inches to 24 inches high. The Anderden Quarries, near Amherstburg, Ontario, were to furnish the stone as had been the case for the enlarged Welland Canal locks. Quarries on nearby Manitoulin Island were to furnish the lock wall backing, consisting of a somewhat inferior quality limestone taken from beds of similar depth. When quarrying commenced at Anderden though, the new beds proved thicker than anticipated, measuring 18 inches, 22 inches, 24 inches and 28 inches thick.[72] Permission was granted to use some courses of 28-inch stone providing the beds of the facing were made proportionately wider.[73]

During the summer of 1890, some 150 men were employed on the Sault Ste. Marie Canal project with work being concentrated on the lockpit excavation, the quarrying of Anderden stone, and the transportation of the stone by barge to St. Mary's Island. Wet weather and strikes for higher wages combined to impede progress somewhat amidst concerns that the contractor had yet to undertake the construction of puddle walls.[74]

To combat potential water seepage problems, clay puddle walls were to be constructed in cut-off trenches. Since the late 18th century, clay puddle walls had been widely used in Britain, and subsequently in the United States and Canada, as a lining to render canal beds and banks impermeable to water where canals passed through sand, gravel, or other permeable ground. It was also widely used as a core wall in earth dams, as well as a cut-off wall behind lock masonry walls.

Solid clay was not suitable for puddling. It would absorb great quantities of water, was permeable by water, and had a disposition to shrink and crack when its water content decreased. On the other hand, clay mixed with a coarse sand or fine gravel and humus to form a light loam, could be made into a clay puddle impermeable to water. To that end, the loam had to be wetted, worked, chopped and kneaded with a spade until it consolidated into a homogeneous plastic material of approximately two-thirds its former volume. The clay puddle so formed was then generally built up, or packed down, in successive layers 10 inches to 12 inches deep to form a watertight wall of the required thickness.

At Sault Ste. Marie, the canal and lock were to be isolated by means of clay puddle cut-off trench walls passing along the entire length of the canal, 50 feet outside of the canal and lock walls. Three cross-walls of clay puddle, connecting the outer puddle wall trenches with the canal, would prevent water flowing along parallel to the canal: one at the head of the canal; one at the upper end of the lock; and one at the lower end of the lock. All of the clay puddle cut-off walls were to be six feet thick, constructed in trenches excavated to the same depth as the canal and lock; and built up in eight-inch-thick layers of moist clay, well-trodden and pounded.

Immediately behind the lock walls, a different type of cut-off wall was planned. It was to consist of poured concrete in the gap between the lock wall masonry and the bedrock in the lockpit, with a five-foot-thick clay puddle wall carried up against the lock masonry above bedrock.

The trench puddle walls were intended to serve two purposes: to aid in preventing ground water seeping into the canal and lock pit excavations during construction, thereby reducing the volume of water needing to be pumped out to enable work to proceed in the dry; and to prevent water escaping from the canal once it was completed and opened.[75]

In commencing work on the lockpit, the contractor had ignored the puddle wall specifications; yet water problems had not materialized. Consequently, on the lockpit excavation reaching its full depth, he had received permission to dispense with the outer trench puddle walls on the section parallel to the lock. The concrete infill and clay puddle cut-off wall to be raised against the back of the lock wall masonry, was considered sufficient to render the lock watertight. The government engineers, however, had continued to insist on the need for the outer trench puddle walls parallel to the canal prism, and had been pushing for their construction.[76]

As of October 1890, a large section of the canal prism had been excavated down to grade. As no difficulty was being experienced in keeping the excavation pumped dry, Ryan asked also to do away with the puddle walls around the canal prism.[77] This was denied, but even more drastic alterations were under consideration in the plan of construction for the Sault Ste. Marie Canal.

"An increase in dimension should be made"

As of the summer of 1890, the CPR had begun to call for a deepening of the Canadian 'Soo' canal to match the depth of the planned 20-foot-deep American navigation system on the upper lakes.[78] Where the lake freighters were concerned, a few feet of additional depth was by no means a trifling factor. The largest of the upper lakes freighters would be able to pass through the 600 foot by 85 foot Canadian lock, with all but a few of the very largest passing two at a time in tandem. Loaded to a draught of 16 feet, they could carry a minimum of 3000 tons. But it was estimated that these vessels could carry up to 5000 tons if loaded to a draught of 20 feet. In effect, over 50 percent more freight could be carried at roughly the same cost per trip. Similar savings could also be effected by increasing the draught for the Welland Canallers.[79] The government canal engineers, however, had advised against any change in the policy of a 16-foot navigation depth,[80] and there matters had rested during the winter of 1890-91.

On 5 March 1891, Sir John A. Macdonald fought his last election on the slogan: "The Old Man, the Old Flag, the Old Policy." The election was a critical struggle in defence of the national policy. It pitted loyalty to the British connection and a transcontinental Canadian economy and transportation system against the Liberal party's policy of unrestricted reciprocity and consequent integration into the

American economy. With the "loyalty" call, and substantial support from Canadian business, new manufacturing concerns fostered by the protective tariff, and the CPR, Macdonald had narrowly averted defeat.[81] Subsequently, his supporters expected their due. In late March, the marine section of the Toronto Board of Trade passed a resolution calling for the deepening of the 'Soo' canal, and the CPR renewed its appeal.[82] With both the Montreal and Toronto business communities pushing for a 20-foot-deep navigation at Sault Ste. Marie, Macdonald overstepped his engineers. Walter Shanly, a prominent Canadian civil engineer, was hired to report on the 'Soo' canal question.[83]

The canal engineers had addressed seemingly all aspects at issue. They had advised that a 20-foot depth of navigation was impractical at the 'Soo' without a corresponding deepening of the Canadian channel in the St. Mary's River below the canal and the dredging of deep water harbours at Port Arthur, Owen Sound, and Port Colborne. The canal excavation could be deepened for about $180 000 but the other works would cost as much as $1 923 000. The major expenditure by far would be incurred in deepening the extensive rocky shoals in the Canadian channel of the St. Mary's River, without which Canadian vessels would still have to pass through American waters to obtain a 20-foot draught below the Sault Ste. Marie Canal.

As matters stood, Canadian trade was being carried for the most part in Welland Canallers and the few Canadian vessels of a larger size and deeper draught would have free access to the planned 20-foot American navigation system. If the Americans should restrict Canadian access, the larger Canadian vessels could be loaded to the lesser draught at the Lakehead to pass through the 16-foot Canadian canal system with the railway carrying a greater proportion of the western grain exports. To a suggestion that the lock alone be built for a 20-foot depth of navigation so as to allow for a future deepening of the canal prism, a technical argument was advanced. A canal deepened beyond the depth of the planned puddle walls would no longer be watertight.

Where shipping on the upper lakes was concerned, no one knew whether vessels had attained their limit for size or draught. This question would not be decided for years to come. Therefore the canal engineers recommended the canal construction proceed as planned. A provision had been made in the original plans for the construction of a second lock, parallel to the first, when traffic demands warranted it. By that time, the questions at issue would be settled, and the new canal lock could be designed accordingly.[84] In sum, it was maintained that:

> the circumstances do not warrant the adoption of a course that would establish such a dangerous precedent as the breaking of an important government contract.[85]

Regardless of such advice, the question had remained open. Walter Shanly's report soon introduced still another dimension.

For the most part, Shanly's report simply reiterated the concerns of the canal engineers as to the implications and costs of deepening the Sault Ste. Marie Canal. He did, however, recommend that the lock chamber be widened. The planned 600 foot by 85 foot lock chamber was designed to pass two of the Welland Canal scale of freighters — vessels 255 feet in length with a 42-foot beam — in a single lockage. The boats were to be placed end to end, with a good deal of room for smaller vessels or tugs alongside. But Shanly pointed out that an increase of only 15 feet in the width of the lock chamber would enable four Welland Canal vessels to be locked through at a time, doubling the capacity of the lock. Simply widening the lock would add no more than $100 000 to the cost. If in future a deeper or larger lock were required, the proposed second lock and a deepened canal prism — presumably with a different system for rendering it watertight — could be constructed.[86]

After assessing the various political, commercial, and engineering considerations at issue, the new Chief Engineer of the Department of Railways and Canals, Toussaint Trudeau, recommended that the dimensions of the Sault Ste. Marie Canal be changed. He proposed that the lock be increased to 650 feet by 100 feet to pass four Welland Canal vessels at a lockage. He also recommended that the lock and canal be deepened to match the 20-foot American canal lock and navigation system already under construction. These changes would cost an additional $225 000 but represented a major saving over the projected cost of constructing a new and larger lock and deepening the canal sometime in the immediate future. Only the enlarged canal had to be put in place at the moment. The other components of a 20-foot Canadian navigation system on the upper lakes — the deepening of Canadian harbours and the provision of a Canadian deep water channel in the lower St. Mary's River — could be postponed until Canadian interests demanded a larger system independent of any potential American control or impositions.[87]

The proposed enlarged lock was authorized by an order-in-council of 21 May 1891 and a supplemental agreement concluded with the contractor, Hugh Ryan & Co. The depth of water on the sills was to be increased from 16 feet three inches to 19 feet for the larger 650 foot by 100 foot lock.[88] In effect, the Canadian canal had to be deepened only two feet nine inches to match the planned 21 feet of water on the sills of the new American Poe lock under construction on the St. Mary's Falls Canal. The American engineers measured their depths from the mean low river level; whereas the Canadians used the lowest recorded river level as their standard.[89] In enlarging the canal, the major projected cost increases were in lock masonry, and here substantial savings were effected through changes in the culvert system.

In the original lock design two sluice culverts were planned. The masonry culverts were to be lined with sheet iron piping of about 10 feet in diameter, and recessed in a trench excavated in the bedrock floor along each side of the lock chamber. The entrance to each sluice culvert was in the lock wall above the upper gates, with both culverts curving downwards through the lock wall and breastwork, and out under the lock chamber floor. Apertures were provided along the top of the culverts to let water in and out of the lock, with hydraulically operated valves controlling its flow through the culverts in filling and discharging the lock chamber.[90]

In designing the enlarged lock, a system of timber culverts was adopted similar to what was in the American Weitzel lock of 1881. A trench, a minimum of 40 feet wide and 10 feet deep, was to be excavated down through the centre of the lock and extended out past the upper and lower gates. In the section of trench extending under the upper gates and through the lock chamber, four timber filling culverts — each 10-foot square in cross-section — were to be constructed, planked over, and blocked off at their lower end. Hydraulic valves placed down in a pit at the head of the culverts controlled the entry of water into the culverts and up into the lock chamber through apertures cut along the top of the culverts. To discharge water out of the lock, a second set of similar timber culverts, but much shorter in length and simply planked solid on top, was to be constructed in the section of trench passing under the lower gates. A short section of trench at either end, both inside and outside of the lower gates, was left uncovered. Hydraulic valves placed in the inner pit would control discharge of water through these culverts (Fig. 8).

The new design adopted for the floor culverts greatly increased the sluice capacity, relative to the size of the enlarged lock chamber, and cut the projected cost of enlarging the Sault Ste. Marie Canal almost in half. An increase of only $100 000 would be required on the estimate, barring any water problems that might be encountered in deepening the excavation. In view of the extra work, the completion date was extended a year to 10 May 1893.[91]

During the summer of 1891, as rock excavation work continued at Sault Ste. Marie, the government acceded to Ryan's request to dispense with the puddle walls along the canal prism. The canal cut was to have been lined with dry stone retaining walls constructed of sandstone from the canal excavation. But the discovery of quicksand in seams of the bedrock being excavated necessitated a change. In response, Ryan agreed to substitute limestone retaining walls laid in cement. This promised to seal the canal against the quicksand as well as make it watertight thereby rendering the puddle walls redundant. The savings effected in doing away with the excavation, pumping out, and building of the trench puddle walls more than compensated for the added cost of cement walls of limestone masonry.[92] While this decision was being made, an even more serious concern arose.

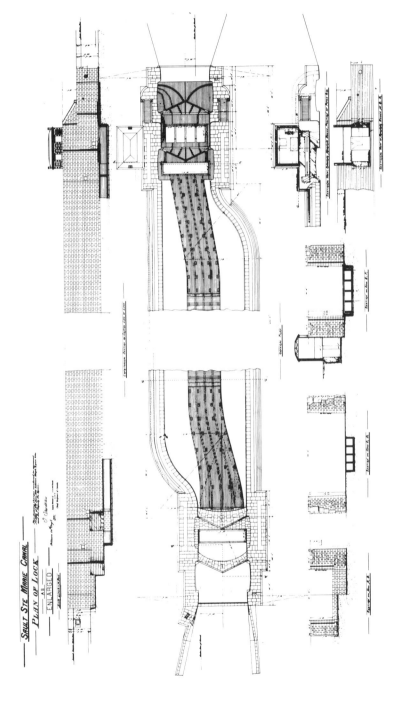

FIGURE 8. Plan of the enlarged 650 foot by 100 foot lock, with off-set gates 60 feet wide, 28 May 1891. (Canadian Parks Service, Sault Ste. Marie Canal Office)

The configuration of the enlarged 650 foot by 100 foot lock conformed to the original lock design in narrowing to a width of 60 feet at the gate openings, with off-set gates. In the wider lock, however, the shoulder of the end walls was much more pronounced curving out 40 feet from the side walls. No sooner had the enlarged lock been adopted than the confined lock entrance was criticized in Parliament.[93] It had become known that a great deal of time was lost by vessels in manoeuvring through the off-set gates of the American Weitzel lock at the 'Soo'. Evidence of the difficulties involved in using a lock of that configuration could be readily seen:

> At both ends the masonry has been scored, chipped, and furrowed by the constant attrition of vessels pocketed at these shoulders, and ground and pounded against the wall in their efforts to gain headway sufficient to make a clean exit.[94]

In the enlarged Canadian flotilla lock, two of the four vessels at any time would be out of the straight channel and compelled to manoeuvre in a very confined space. Where speed and ease of passage was concerned, the advantages of a straight lock could not be denied.

The Novel Long Lock

Initially, some consideration was given to simply widening the gates of the enlarged lock.[95] The new 800 foot by 100 foot Poe lock was being constructed on the American St. Mary's Falls Canal as a straight lock with 100-foot gate openings and arched steel mitre gates. Each gate leaf was 56 feet wide and, in the case of the lower gates, 43 feet high weighing 157.5 tons.[96] The Canadian engineers were confident that such massive gates could be effectively operated by hydraulic power, but remained apprehensive about the dangers implicit in any accident to lock gates 100 feet wide. Replacement gates would also be very costly, and require a great deal of time to construct.

After further consideration, it appeared that the advantages of a straight lock could best be attained through altering the configuration of the lock. If the lock chamber was lengthened and reduced in width, vessels could lock through one astern of another, rather than being confined two abreast in two rows as in the enlarged lock with off-set gates. Hence, three proposals were submitted to Mackenzie Bowell, the Acting Minister of Railways and Canals, in keeping with the novel long lock configuration. Two called for a lock designed strictly for Welland Canal vessels: a 1100 foot by 60 foot lock capable of passing four of the Welland Canallers in a single lockage; and, what the Chief Engineer of Canals preferred, an 830 foot by 60 foot lock to pass three of these vessels at a time. The third proposal was for a lock 900 feet by 60 feet to pass two Welland canallers and a single 320-foot upper lakes vessel in a single lockage. In all cases, the gates were to be 60 feet wide and constructed of wood, with 19 feet of water on the sills of the lock (Fig. 9).[97]

Bowell preferred the 900 foot by 60 foot lock, and an order-in-council of 24 December 1891 authorized its construction, pending negotiations with the contractor Hugh Ryan & Co.[98] The novel configuration of the proposed Canadian long lock, however, did not appeal to shippers operating steam barges on the upper lakes. They much preferred the configuration of the 800 foot by 100 foot American Poe lock. When completed it would handle four of the largest steam barges; whereas the 900 foot by 60 foot Canadian lock would not hold three of these vessels. The more traditional wide configuration of the American flotilla lock also was judged better suited for passing lumber tows, tugs, and passenger steamers. Where speed of lockage was concerned, ship captains maintained that in a large straight lock steam barges could be put in alongside one another just as quickly as stopped close up one astern of another in the long lock.[99] A good deal of criticism was also levelled at the 60-foot width of the projected Canadian lock which "while far too wide for a single boat, is much too narrow for two boats side by side, so that a very considerable space is useless."[100] Whether judged in terms of the Canadian Welland Canallers, the large CPR passenger steamers or the American upper lakes freighters, the projected width of the Canadian lock appeared completely arbitrary.

Where the Canadian ship canals were concerned, vessels had been increasing dramatically in length and draught, but comparatively little in breadth of beam. The Lachine Canal of the 1840s canal building program had had locks 200 feet by 45 feet with nine feet of water on the sills. In the on-going canal enlargement program of the 1870s, the locks were constructed 270 feet by 45 feet with initially 12 feet, and then 14 feet of water on the sills. In effect, after 30 years, the width of the lock chamber had not required any increase to accommodate Canadian lake vessels, but the length was increased by a third and the depth by a half.[101] A similar situation pertained to upper lakes shipping.

On the Sault Ste. Marie Canal project planning had proceeded hitherto in keeping with the scale of the enlarged Welland and St. Lawrence canals. The various dimensions adopted for the huge flotilla lock were based on multiples of the 255 foot by 42 foot Welland Canallers, and the increase to a 16-foot and then 19-foot depth was made simply to augment the loading capacity of these vessels on the upper lakes leg of their voyage. The CPR passengers steamers, although of greater length, were of about equal or lesser beam. The two original passenger steamers remaining in service, the *Algoma* having been lost in a storm, were 263 feet long with a 38-foot beam. The newest CPR passenger steamer, the *Manitoba* launched in Owen Sound at the new Polson Iron works in 1889, was 303 feet long with a 43-foot beam.[102] None of these Canadian vessels required a 60-foot- wide lock chamber; nor did the American steam barges proliferating on the upper lakes.

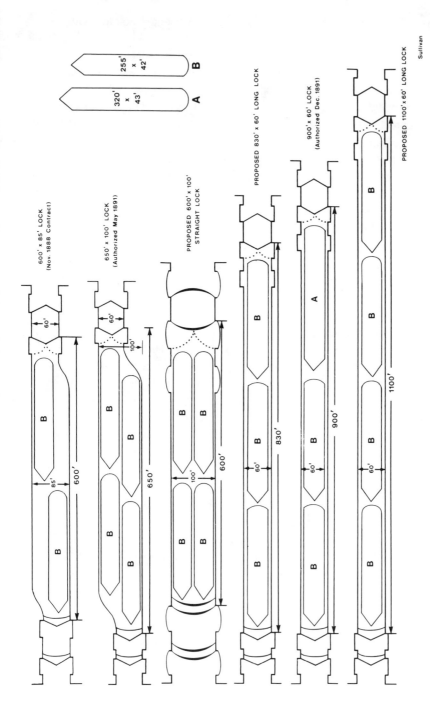

FIGURE 9. Outline of lock chambers considered, or authorized, for construction between November 1888 and December 1891. (Canadian Parks Service, 1986)

The vessels designed specifically for the upper lakes trade had been increasing greatly in length proportional to their width, from a 6:1 ratio — the proportions of Noah's Ark — to over 7:1, and beyond almost to 8:1. With few exceptions, as of 1891 the largest of the steam barges were about 320 feet long. They had gained 100 feet in length in just over two decades; yet remained less than 43 feet in beam (Fig. 10).[103] The newest type of lake freighter, the whaleback introduced in 1888, had attained 265 feet in length with a 38-foot beam or less, and the sailing schooners were somewhat smaller but similarly proportioned.[104] Existing boat dimensions

FIGURE 10. The S.S. *B.W. Arnold*, a typical large lake freighter of the 1890s with two schooners in tow leaving the Weitzel Lock, St. Mary's Falls Ship Canal. (National Archives of Canada, PA-139088)

did not justify a 60-foot-wide lock. Nonetheless, a decision had been taken. It remained to be seen whether the novel long lock configuration would prove as proficient as its designers predicted.

Lock Masonry Work

As of 5 April 1892, detailed working plans were completed for the masonry of the 900 foot by 60 foot lock and a second supplemental agreement concluded with Hugh Ryan & Co. The completion date for the 'Soo' canal project was extended to 31 December 1894 and the estimate increased from $3 million for the enlarged 650 foot by 100 foot lock with off-set gates, to $4 million for the long straight lock.[105] The agreement maintained the existing specifications in force, but slightly altered to conform to the new lock dimensions. The masonry walls, 44 feet six inches high in the lock chamber, were now 1106 feet long overall and 20 feet four inches thick at the base narrowing to 10 feet 11 inches at the top, with a one-inch in 48-inch batter on the water face and step-backs at the rear. Around the gates, to take the thrust of the water pressure acting against each mitre gate leaf, the wall masonry thickened to 25 feet six inches at the base and carried straight up to the lock wall coping at that thickness (Fig. 11).

Out of a total of 75 000 cubic yards of masonry in the lock, approximately one-third consisted of cut stone and the remainder of backing stone.[106] The cut stone facing would consist of the Anderden limestone laid, with headers and stretchers, in regular courses 18 inches, 22 inches, 24 inches and 28 inches in depth. The backing wall was to be constructed of rough squared blocks of Manitoulin limestone intermixed with usable sandstone from the canal excavation. The backing courses were to be generally of the same depth as the corresponding face stone, but occasionally in the heavier courses two thicknesses of backing stone were acceptable providing the top of such courses was on the same level as the corresponding course of the heavy face stone. This, of course, was necessary to ensure that the headers tying the face stone into the backing were properly laid on a level bed. The backing stones were to be laid in Canadian hydraulic cement, mixed with clean sharp sand. Each stone was to be thoroughly bedded in mortar with no contact of stone against stone and no pinning allowed.[107] The cut stone facing was to be laid in Canadian natural or so-called "native" cement, and pointed or "lipped" with Portland cement.[108]

In canal works, the mortar used in laying stone masonry was invariably made with an hydraulic cement. It would harden under water or in damp conditions where common limestone mortar simply crumbled or washed away. Both natural and Portland cements had hydraulic properties, but differed greatly in their manufacture and customary usage.[109]

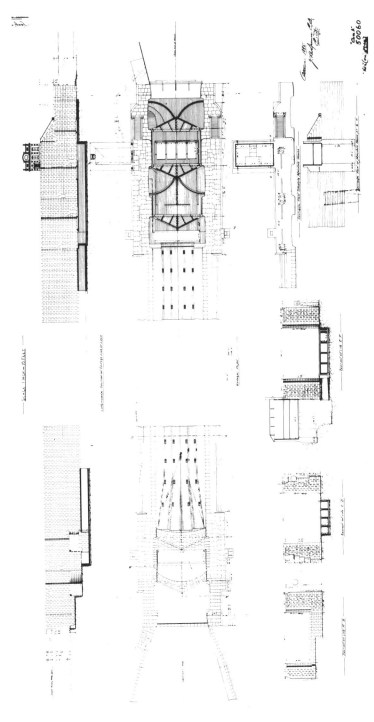

FIGURE 11. Plan of lock as constructed, Sault Ste. Marie Canal, 1895. Contrary to the drawing, the lock walls were constructed with a rubble concrete core rather than cut stone masonry. (Canadian Parks Service, Sault Ste. Marie Canal Office)

Natural cements were manufactured from naturally occurring hydraulic limes or limestone containing clay, chalk or marl, calcinated in a kiln, and ground into a powder. The natural hydraulic limes, however, varied greatly in their composition, properties and presence of impurities, and were considered less than totally trustworthy in quality and behaviour.[110]

Portland cement in contrast was an artificial cement manufactured by adding argillaceous earth or clay to common limestone in set proportions, and calcining the mixture in high temperature kilns. From the time of its first manufacture in England in 1825, Portland cement had proved far superior to natural cements. It had a uniform high quality, set faster, and dried harder; but had not come into widespread use. Portland cement was highly expensive to manufacture, and far too costly for use in heavy masonry work. It had remained so until the introduction in 1889 of the rotary kiln, using a new continuous feed wet process of manufacture, brought prices down to more acceptable levels.[111] Delivered at Sault Ste. Marie, Portland cement cost about 80 percent more than the equivalent of Canadian natural cement[112] — a price differential reflected in the lock masonry specifications governing their respective use.

Imported English natural cement had been used whenever hydraulic cement was required for Canadian canal construction work until the discovery of an hydraulic lime near Hull, Quebec, in June 1829. The Hull natural cement, manufactured by Philemon Wright & Sons, had proved a reliable substitute when used to construct the lock masonry of the Rideau Canal.[113] Subsequently natural hydraulic cement from Canadian plants established at Kingston, Queenston, and Thorold, had been used in building the masonry of the second Welland Canal, the St. Lawrence canals of the 1840s, and in the ongoing canal enlargements of the 1870s and 1880s. On the commencement of the Sault Ste. Marie Canal project in 1888, the Portland cement used in Canadian canal construction was imported from England. But as of 1891, several Canadian Portland cement plants were being established using the new rotary kiln technology, opening up the possibility of using a Canadian supplier.[114]

On the 'Soo' lock, Portland cement concrete was to be used to a limited extent in critical areas around the foundation of the masonry lock walls. It was to be used only as an infill, primarily to seal the gap between the lock masonry and the rock walls of the lock pit excavation. Above bedrock a three-foot-thick clay puddle wall, built up against the rear of the lock wall, would prevent the passage of ground water. Otherwise, puddle walls were discarded. Portland cement concrete was also to be used in levelling up the bedrock floor of the lock chamber and for the same purpose in the culverts trench down the centre of the lock (Fig. 11).[115] The timber floor culverts system, introduced earlier for the proposed 850 foot by 100 foot enlarged lock, was retained.

Floor Culverts

Where the floor culvert system was concerned, Canadian engineers were strongly influenced by the design features of the American Weitzel lock of 1881, and the new Poe lock on which construction was proceeding. Indeed, the layout and construction details of the timber floor culverts of the proposed new Canadian long lock mirrored the earlier American design down to even the type of valve employed in the culverts.

In the Canadian lock plan, the trench for the filling and discharge culverts ran down the centre of the lock. It was 11 feet deep, below the bedrock floor of the lock chamber, and extended from just outside the upper gates through to just beyond the lower gates. The width varied somewhat: 61 feet wide at the upper gates, narrowing down to 42 feet through the lock chamber, and 61 feet wide throughout at the lower gates (Fig. 11).

In the trench a number of heavy longitudinal sill timbers, 12 inches by 15 inches, were anchored to bedrock with 1.25-inch rock bolts, six feet long at six-foot spacings. Transverse timbers, 12 inch by 12 inch, were laid across the sills six inches apart and the interstices filled with Portland cement concrete and grouted flush with the top. Two layers of planking, two and three inches thick, formed the floor on which the side walls of the four filling and four discharge culverts rested. These walls were two feet thick and consisted of two thickness of 12 inch by 12 inch timbers. The solid timber culvert walls supported the transverse floor beams of the lock chamber — 12 inch by 12 inch timbers placed six inches apart — and two layers of planking: a two-inch plank floor over a three-inch plank sub-floor. Iron straps were the placed across the planks over the heads of long rods, spaced two feet apart and passing down through the timber culvert walls to the longitudinal sills anchored to bedrock. Nuts tightened down on the threaded vertical rods would keep the floor from heaving up. At the entrance to the filling culverts, and at both ends of the discharge culverts, a well was left open for the passage of water. These three wells extended across the full width of the culvert trench.

The filling and discharge culverts differed slightly. The four discharge culverts were each a uniform eight feet high by 10 feet six inches wide; whereas only the entrance to the filling culverts was of that size. After passing under the breastwall, each filling culvert narrowed to a uniform eight feet square through the lock chamber. As many as 152 apertures were to be cut in the plank flooring over the filling culverts to let water into the lock chamber.[116]

A single butterfly valve of steel — approximately eight feet by 10 feet, and mounted on a horizontal axis — was placed in the entrance to each filling and discharge culvert. On the American locks, the huge butterfly valves were worked by an hydraulic cylinder in the sluice culvert well. Each valve was pivoted open and closed by a single cylinder with a connecting rod attached directly to one arm of the valve plate (Figs. 12 and 13).[117] In the Canadian design, as initially proposed for the abortive 600 foot by 85 foot lock, the butterfly valves were to be operated by a system of cables and snub pulleys, with two cables attached to each valve plate to pivot it open and closed. The cables passed up out of the sluice culvert well, across the lock floor, and up wells inside the lock wall masonry to hydraulically operated sliding pulley machines on the lock wall coping (Fig. 14).

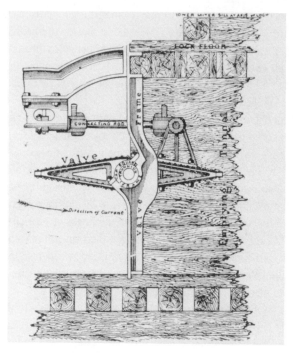

FIGURE 12. Butterfly valve system on the 1881 Weitzel lock, showing the direct-connected piston of the hydraulic cylinder in the discharge sluice culverts. (*Engineering and Mining Journal*, 30 August 1890)

In this plan, the sluice culverts were to work in tandem, with one valve machine operating two of the adjacent butterfly valves. In effect, four machines were needed: two to work the four filling culvert valves; and two to work the four discharge culvert valves. A set of two hydraulic lines — a pressure pipe and an exhaust pipe return — ran from the powerhouse to the valve machines on the lock coping. On the powerhouse side of the lock, the pipes ran outside the lockwall masonry at a depth of eight feet six inches, well below the frost line. The lines to the machines on the other side of the lock ran down through the lock wall masonry, across under the wooden floor, and up through the lock wall on the other side (Fig. 11).

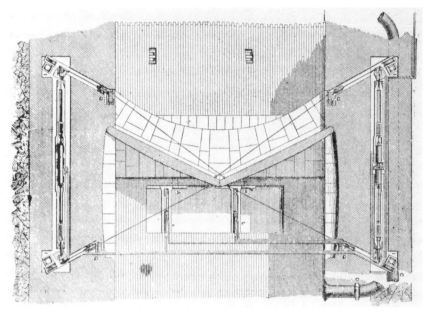

FIGURE 13. Hydraulic cylinders on lock walls, for operating gate leaf cables of the American Weitzel lock. The hydraulic cylinders for operating the two sluice culvert butterfly valves are in the filling well below the gates. (Engineering and Mining Journal, 30 August 1890)

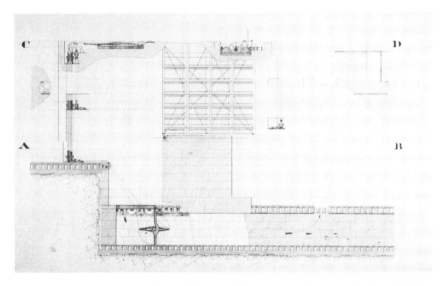

FIGURE 14. Proposed lock machinery, 15 January 1891. A system of cables and pulleys initially proposed for working the butterfly valves hydraulically on the 600 foot by 85 foot lock. (Canadian Parks Service, Sault St. Marie Canal Office)

Why the Canadian engineers initially chose to depart from the American layout with its simple direct connection between the hydraulic cylinder piston and the valve plate, is unclear. The American system was simplicity itself, compact and effectively protected from injury in the culvert wells. Yet it was ignored initially in favour of a complex system of cables and pulleys exposed on the lock floor to potential fouling with debris and to being damaged, if not torn away, by heavily laden boats. Placing the hydraulically operated sliding pulley machinery up on the lock wall coping no doubt would facilitate access for repairs to that component of the system, but the amount of hydraulic piping required was much greater than in the American system. The more numerous moving parts also were far more susceptible to wear and break down. Ultimately, however, the superiority of the American hydraulic valve system was recognized and adopted for the planned 900 foot by 60 foot lock.

The Canadian canal engineers not only adopted the American system of direct-connected hydraulic cylinders, but improved on it. The American layout, with the hydraulic cylinders positioned in the valve pits and the piston directly attached to each valve plate, was abandoned. It would have required eight cylinders, one for each of the four filling and the four discharge sluice culvert valves.

In the new Canadian layout, only four cylinders were required, and each was to be mounted on the lock wall coping to operate a vertical draw-rod 55 feet long. The draw-rod was positioned in a well passing down through the lock wall masonry into a small chamber where the rod was attached, by a slotted connection, to a four-foot-long crank arm on the extension of the horizontal shaft on which the floor sluice culvert valves pivoted. Two butterfly valves were mounted on each shaft, so as to operate in tandem. Each hydraulic piston in lifting the draw rod through a travel of five feet six inches would rotate two valves through an arc of 90 degrees to the fully open position, and in lowering the draw-rod would pivot the two valves closed (Fig. 32).[118]

Lock Gates

The mitre gate leafs of the Canadian lock were of a standard trussed bowstring construction, framed white oak timbers sheathed with pine planking on the curved upstream face (Fig. 15).[119] The size and number of the gates, however, was effected by two design changes derived from the American Poe lock.

Traditionally locks opening up into major rivers or large bodies of water, were equipped with four sets of gates — the upper and lower gates for operating the lock, and upper and lower guard gates. The reasoning was that canals could not be easily drained to replace or repair lock gates; hence guard gates were provided for clos-

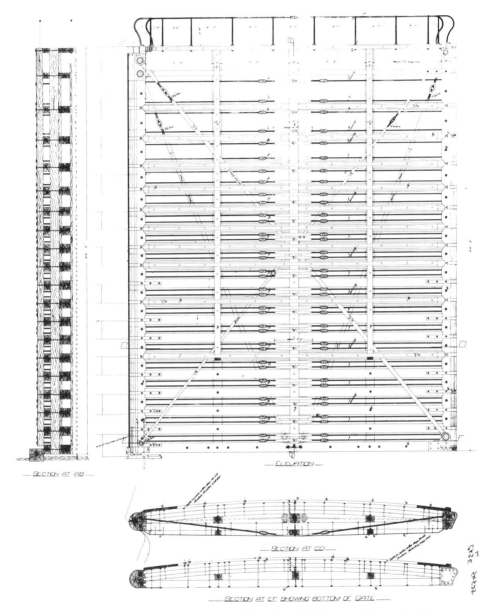

FIGURE 15. Plan of upper main spare gates, 10 December 1894. The trussed bowstring lock gates initially erected on the Canadian lock. (Canadian Parks Service, Sault Ste. Marie Canal Office)

ing off the canal to dewater the lock. Large ship canal locks were almost invariably constructed with guard gates, but on the Poe lock a set of intermediate lower gates had been added. These gates were identical to the lower gates, and were intended for use in the event the lower gates required repairs rendering them temporarily inoperable. A similar set of auxiliary gates was also incorporated into the Canadian lock and all of the gates lengthened somewhat to take account of another innovation (Fig. 11).120

As initially planned, the lower sill was to be constructed on the lock floor in keeping with traditional lock building practice. The Americans, however, had recently increased the projected depth of draught on their canal by one foot through setting the lower sill of the Poe lock down into a floor recess. To maintain an equivalent depth, the sills of the Canadian lock were also lowered, obtaining a 20 foot three inch depth of water on the sills at extreme low water level.[121]

With the lowering of the sills, the largest gates on the Canadian lock — the lower and auxiliary gate leafs — were 44 feet six inches high and 37 feet wide, weighing 87 tons apiece. These gates would be by far the largest lock gates ever constructed on a Canadian canal; but by no means as huge as the arched steel gates on the projected Poe lock.[122]

Each gate leaf was to operate by means of cables attached to the front and rear of the mitre post near its base. The cables ran across above the floor to a horizontal pulley in the lock wall masonry, and up inside an inclined well to the operating machinery on the wall coping. A gate machine — six in all, positioned at the upper, intermediate, and lower gates — operated the opening cable of the adjacent gate leaf, and the closing cable of the gate leaf on the opposite side of the lock. This cable arrangement had long been used on Canadian canal locks with manually operated crabs for working the cables. At Sault Ste. Marie, the engineers simply planned initially to substitute an hydraulically operated gate machine with winding drums for the crab (Figs. 14 and 16). The older manually operated crabs, however, were to be retained for the guard gates.[123]

The Americans on the Weitzel lock of 1881 had employed the conventional cable arrangement for operating the mitre gates, but worked the cables by means of two hydraulic cylinders placed back to back on the lock coping at each gate leaf. The cables were attached directly to the pistons of the hydraulic cylinders. By the time planning proceeded on the 900-foot lock, the hydraulically operated winding drums were discarded in favour of adopting the far simpler American gate operating machinery (Fig. 13).[124]

During the summer of 1892, work proceeded in excavating the trench for the floor culverts of the lock and in deepening the canal prism. Then on 15 September 1892, the first stone of the lock masonry was laid,[125] as negotiations were undertaken to speed the construction of the Sault Ste. Marie Canal.

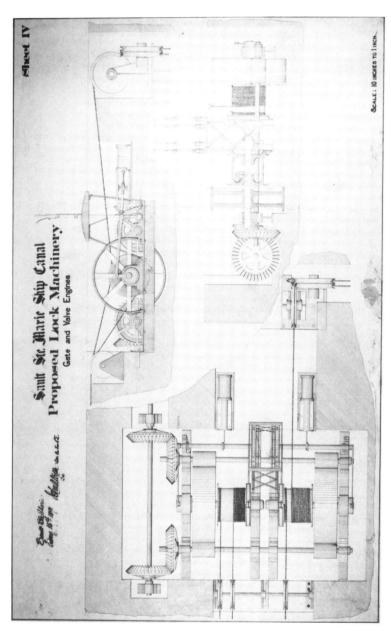

FIGURE 16. Proposed lock machinery, gate engines, 15 January 1891. Hydraulically operated winding drums were initially proposed for working the lock gates cables of the planned 600 foot by 85 foot lock. (Canadian Parks Service, Sault Ste. Marie Canal Office)

"if concrete were substituted"

Despite American protests, the Conservative government had kept its toll rebate system in force on the Welland-St. Lawrence canals system to encourage shippers to use Canadian ports for their export trade. An order-in-council of March 1892 had set the Welland Canal toll at 20 cents a ton with vessels proceeding to Canadian ports to receive a rebate of 18 cents a ton. The American government had renewed its earlier protest and retaliated with the Curtis Act. If tolls on American commerce were not removed by 1 August 1892, a toll of 20 cents a ton would be imposed on all ships passing through the St. Mary's Falls Ship Canal in transit to Canadian ports.[126] The critical importance of an all-Canadian transportation system — the keystone of the Conservative party's national policy — was now evident to all concerned.[127] Almost immediately government engineers met with Hugh Ryan & Co. to discuss ways of bringing the Sault Ste. Marie Canal project to a rapid completion.

On 8 November 1892, a new agreement was signed. Ryan agreed to have the lock walls finished as of 1 December 1893, and the timber floor culverts completed as of 1 July 1894 together with the canal prism, which was to be deepened to 22 feet in keeping with the earlier decision to lower the lock sills. Ryan also contracted to construct the masonry pier and abutments for a railway swing bridge by the same deadline. Earlier in 1887, the CPR had constructed a branch line across the St. Mary's River to tap into the railway system of Michigan's upper peninsula. The swing bridge would replace a trestle carrying the railway across the canal excavation.

The new agreement advanced the completion date six months, but would result in the canal being in operation almost a complete navigation season earlier than previously planned. On its part, the government agreed to pay a $90 000 bonus to cover the extra costs involved in speeding up the work. Hugh Ryan & Co. were also authorized to substitute Portland cement concrete for the stone masonry backing of the lock chamber walls between the upper and lower gate piers, if deemed necessary to meet the new schedule.[128]

As of 1892, concrete had not been used as a structural component of any Canadian lock; but in England the locks on the new Manchester Ship Canal were being constructed with concrete backing walls faced with brick. Moreover, the Department of Railways and Canal had already decided, after studying the Manchester Ship Canal plans, to construct the new Soulanges Canal with concrete locks faced with cut stone. The Soulanges Canal, on which excavation work was just commencing, was to replace the old Beauharnois Canal as part of the ongoing St. Lawrence canals enlargement.[129] This novel use of concrete would speed construction and cut masonry costs. Indeed, the government engineers calculated that

$14 518 would be saved if the backing of the 'Soo' lock were constructed completely in mass concrete between the gate piers.[130] The contractor, however, had a different system in mind.

During the summer of 1893, the plant and workforce on the Sault Ste. Marie Canal project was nearly doubled to compensate for the loss of a work season.[131] Additional steam-powered derricks were constructed (Figs. 17 and 25), totalling 40 in use on the project, and steam shovels were put to work in the canal excavation. Ten miles of track were laid, giving the bottom of the canal the appearance of a railway yard, and as many as 53 tram cars drawn by teams of horses were put in service and switched over the track system carrying stone and concrete.

FIGURE 17. Building the Canadian lock. While excavation work proceeds, workmen mix concrete by hand on the platform where horse carts are dumping the aggregate. (Canadian Parks Service, Harry Ross Collection)

In the lock pit excavation, Hugh Ryan & Co. built two massive travelling derricks for raising the lock walls. These derricks travelled on a railway track of 48 foot six inch gauge, and spanned the 12-foot-deep culvert trench running down the centre of the lock. Four masts were mounted on each travelling derrick, two for each lock wall, with steam power driving both the hoisting cable drums and the drive wheels for moving the derrick (Figs. 18 and 20). A second stone crusher was also added. It augmented the existing crusher, worked by cable from a turbine in Ryan's powerhouse. One crusher had a capacity of 30 tons per hour and the other 40 tons per hour.[132]

In keeping with the accelerated work schedule, the workforce was increased to a total of over 800 men. By far the largest component, 400 men, was employed in excavating the canal prism. Another 200 worked in the Anderden and Manitoulin Island quarries getting out thousands of tons of stone which kept a fleet of tugs and barges busy providing transport to St. Mary's Island. At the lockpit, 200 stone cutters and masons prepared the stone and laid up the wall masonry, with a large number of mechanics, carpenters and 18 blacksmiths providing support through keeping the machinery and tools in repair (Fig. 19).[133] To further speed construction, night work was undertaken with light being provided through burning the wooden cement barrels.[134]

Prior to the spring of 1893, work had been undertaken on only the north wall of the lock in keeping with the original specifications. The contractor immediately abandoned this approach, exercising his option to employ concrete in the lock wall backing. Rather than pour a solid mass concrete backing wall, however, Ryan had proceeded with a system of his own devising.[135]

In laying up the lock walls, each course of face stone was set together with the corresponding face course of backing stone, of equal course depth, in the rear wall. Large slabs of stone, taken from the natural deep beds of sandstone being excavated from the canal prism, were put in the wall cavity, keeping clear of the cut stone in the wall faces. The irregularly shaped flat stones were bedded on either Portland cement mortar or an infill of Portland cement concrete. Concrete was then placed around the stone slabs, and the protruding headers of the face stone. In this manner a solid masonry wall was formed, with the concrete infill levelled up across each course of stone masonry as it was raised on the wall faces and grouted (Figs. 20, 21 and 22). Ryan, in effect, constructed a concrete core wall but one of rubble concrete built of large slabs of stone embedded in concrete rather than a poured core of plain concrete.[136] Both limestone from the Manitoulin quarries and sandstone from the canal excavation, were used in constructing the backing walls.

This system greatly speeded construction of the lock wall backing as the interior stone did not have to be bush-hammered and rough-squared as would have been the case had the backing wall been laid in solid stone masonry with tight mortar joints and level beds. It also yielded several advantages where building materials were concerned. A large quantity of four- to nine-ton blocks of Manitoulin limestone had been quarried already and transported to the site for use in constructing the backing according to the original specifications. Rather than being discarded, these limestone blocks were used in laying up the rear face of the backing wall.

FIGURE 18. Building lock walls. A travelling derrick straddles the floor culverts trench in the lock chamber. (Canadian Parks Service, Harry Ross Collection)

FIGURE 19. Stone masons at work, preparing blocks of cut stone for the lock walls. (National Archives of Canada, C-7651)

FIGURE 20. Building lock walls. A four-masted travelling derrick used in constructing the heavy stone masonry lock walls, 1893. (Canadian Parks Service, Sault Ste. Marie Canal Office)

FIGURE 21. Lock gate anchor set in the lock wall among the irregular slabs of stone later infilled with concrete, 1893. (Canadian Parks Service, Sault Ste. Marie Canal Office)

FIGURE 22. Building the breastwall at the upper gates. The lock wall backing at the breastwall reveals how the irregular slabs of stone were laid with thick joints of concrete, 1893. (Canadian Parks Service, Sault Ste. Marie Canal Office)

Likewise, the irregular slabs of stone being removed from the natural beds of sandstone in the canal excavation, could be used. Embedded in concrete, they reduced significantly both the amount of Manitoulin limestone that had to be quarried and shipped to the worksite as well as the volume of concrete otherwise required to construct the lock wall backing.

All of the concrete was made of Portland cement mixed with sand and broken stone chips from the canal excavation and the stone cutting yards.[137] At first, imported English Portland cement was used. But thereafter Canadian Portland cement was purchased from the new "Owen Sound Portland Cement Works," and found of equal quality with the best English cement.[138]

Both the cut stone face of the lock walls, and the backing stone face at the rear, were laid initially in Canadian natural cement, with the mortar joints of the face stone being pointed in imported English Portland cement. Some 37 686 barrels of natural cement were procured from Battle & Sons and Usher & Co. of Thorold, Ontario. Sometime during the summer of 1893, however, a change was made. The Chief Engineer, Department of Railways and Canals, directed Portland cement to be substituted for natural cement in laying up the cut stone facing and the backing wall face stone. Subsequently about 28 000 barrels of Portland cement were used to complete the stonework,[139] with an equally impressive amount of Portland cement being used in making concrete.

About 24 000 cu. yards of concrete were put in the Sault Ste. Marie lock chamber, including the concrete used in levelling up the bedrock floor and lining both sides of the culvert trench rock excavation with a concrete wall (Fig. 23).[140] Although this was not a solid mass of concrete placed in a continuous pour, it was nonetheless an impressive volume of concrete in contemporary terms.[141] Within five months, almost 63 000 cu. yards of masonry — both stone and concrete — were placed in the Sault Ste. Marie lock. This, the engineers claimed, was "more masonry ... than has before been built in one structure in the same time." Indeed, all of the lock masonry work was completed as of 16 November 1893, six weeks ahead of the accelerated schedule (Figs. 24 and 35).[142]

On completion, the walls of the Sault Ste. Marie Canal lock were regarded by many Canadian engineers and stone masons as being the finest example of lock masonry to be found anywhere. Although none had ever heard of lock walls constructed in keeping with the hybrid system introduced by Hugh Ryan & Co., they were unanimous in regarding the introduction of concrete as a decided improvement. The rubble concrete core wall, and the use of Portland cement throughout, rendered the lock far more watertight than would have been the case otherwise.[143] Indeed, the masonry of the Canadian lock was considered far superior to the American Poe lock which was being constructed of solid stone masonry in the con-

FIGURE 23. Timber floor sluice culverts under construction, 1894. A concrete retaining wall was poured along both sides of the trench excavated for the culverts. (Canadian Parks Service, Canal Records, File C-4272/S32, Vol. 9)

FIGURE 24. Upper wing walls of the lock, 21 September 1894. The penstock intake for the power-house turbines is visible at left. (Canadian Parks Service, Canal Records, File C-4272/S32, Vol. 9)

ventional manner. There both the cut stone facing and backing stone were being laid in native cement and the face merely pointed with Portland cement.[144]

With watertight masonry, clay puddle walls were no longer considered necessary to seal off the Canadian lock. The backfilling was done simply by using hydraulic pressure. A 25-foot- to 30-foot-deep mass of sludge was flashed or washed in behind the lock walls and packed down without resorting to the more conventional tamping.[145] Work on the Sault Ste. Marie Canal was speeded up still further through adopting a different method of construction, utilizing materials on site, for the retaining walls of the canal prism.

When work commenced in deepening the canal to 22 feet below the extreme low water level, the retaining walls were to be constructed of grey limestone laid in Portland cement. On reaching the full depth of excavation, however, the bedrock in some sections had proved "of too shaky a character" to support a masonry wall. Consequently, the canal excavation was widened in the unsound rock and rock-filled timber cribs constructed. The cribwork was to be carried up to a height of 21 feet six inches or within six inches of the extreme low water level. On this foundation, the masonry retaining wall of randomed course limestone would be raised another 10 feet to the height of the surrounding land. Where the bedrock was sound, the limestone wall was to be raised the full height of the canal prism.[146]

FIGURE 25. Timber cribwork under construction along the canal prism, 26 November 1893. (Canadian Parks Service, Sault Ste. Marie Canal Office)

Construction proceeded rapidly during the winter of 1893-94, with but minor modifications. Cribwork was erected along almost the entire length of both sides of the canal cut (Fig. 25). But in some places, it was built along the top of the natural rock face of the canal excavation and carried up only three feet or more in height to form a level foundation for the stone masonry to be erected above. Forcing work during the winter months, however, ultimately proved unwise. In May 1894, several sections of the cribwork pushed outwards anywhere from two to three feet. In filling the cribs, a good deal of frozen clay had become intermixed with broken rock taken from the canal excavation and prevented the rock from compacting properly. In contrast, the loose stone fill behind the cribs had been thoroughly compacted by the weight of huge piles of spoil taken from the canal excavation and by the derricks moving along the outside of the wall in constructing the cribwork. On the ground thawing, the backfill had compacted even further under the weight of the derricks forcing the several sections of cribwork outwards.[147]

To remedy the situation, a total of about 205 linear feet of cribwork, and 282 linear feet of masonry retaining wall resting on the cribwork, was taken down. In these sections, the excavation was cut back and the cribwork reconstructed with a 10-foot wider base. To strengthen sections where the cribwork had been forced forward so as to overhang the natural rock wall of the canal excavation, temporary timber braces were used to prop up the outer face of the cribs. These braces which sloped outwards anywhere from two to six feet at their base, were then planked to form cribbing, and filled section by section with pours of Portland cement concrete carried up to the base of the cribwork (Figs. 26, 27 and 28). This was the first use of plain concrete in a retaining wall on a Canadian ship canal and reflected growing familiarity and confidence in concrete as a building material. It also saved a great deal of time and expense. Lacking this expedient, all of the cribwork would have had to be torn down and reconstructed.[148]

On commencing the masonry retaining wall, the contractor had received permission to use sandstone from the canal excavation. This eliminated the need for further shipments of limestone from the Manitoulin Island quarries, and ultimately only a quarter of the retaining wall masonry was laid in limestone. The rest consisted of large coarse sandstone blocks, dressed only on the beds and ends to make tight mortar joints of random coursed masonry. The masonry walls varied greatly in height along the canal prism, depending on the height of the cribwork foundation beneath, but were constructed with a uniform 2.5-inch to 12-inch batter on their water face (Fig. 28).[149]

On the Sault Ste. Marie Canal project, a great deal of work was accomplished by the scheduled completion date. As of 30 June 1894 the canal prism walls were half built. Otherwise all masonry work was completed including the three-storey powerhouse and the substructure of the railway swing bridge on which the superstructure was already in place (Fig. 29). In the lock chamber the timber culvert sluices were finished, and work was nearing completion on the five pairs of lock gates as well as the eight steel butterfly valves for the floor culvert sluices (Figs. 30, 31 and 32). In addition, work was well advanced on the contracts for constructing a six foot eight inch-diameter steel penstock leading to the powerhouse (Fig. 33), and for the furnishing of two turbines for the powerhouse as well as two centrifugal pumps which were to be operated by the powerhouse turbines and used in dewatering the lock chamber.

Despite a prodigious effort, the canal was not completed in keeping with the accelerated schedule. Work continued throughout the summer of 1894 on the canal prism walls and on installing the operating machinery which was all of Canadian manufacture.[150]

FIGURE 26. Removing section of timber cribwork to reconstruct the canal wall on a wider base, 18 August 1894. (Canadian Parks Service, Sault Ste Marie Canal Office)

FIGURE 27. Temporary props supporting cribwork forced out several feet beyond the rock face of the excavation below, spring 1894. (Canadian Parks Service, Sault Ste. Marie Canal Office)

FIGURE 28. Canal wall reconstruction, 1894. Concrete retaining wall poured up in sections beneath the overhanging timber cribwork. A randomed course masonry wall was erected above the cribwork. (Canadian Parks Service, Sault Ste. Marie Canal Office)

FIGURE 29. Timber cribwork lining lower canal entrance, with gates on lock floor ready for erection, September 1894. (Canadian Parks Service, Sault Ste. Marie Canal Office)

FIGURE 30. The new lock chamber, September 1894. The outlet for the four discharge culverts is in the foreground. (Canadian Parks Service, Sault Ste. Marie Canal Office)

FIGURE 31. The four steel butterfly valves in the filling well, 1894. The culverts pass through the masonry breastwall, below the upper gates, and out under the floor of the lock chamber. (Canadian Parks Service, Sault Ste. Marie Canal Office)

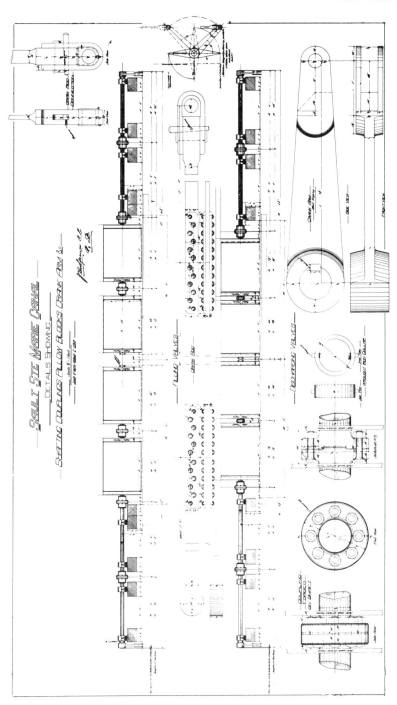

FIGURE 32. Shafting, crank and draw-rod of valve system, 1894. The crank arm, on the end of each shaft, worked two butterfly valves in unison. The crank arm was pivoted by the vertical draw-rod attached to a vertical screw power machine on the lock wall coping. (Canadian Parks Service, Sault Ste. Marie Canal Office)

FIGURE 33. Laying the steel penstock along the lock wall from the intake to the powerhouse, 1894. Wooden saddles supported the sections of pipe until bedded in concrete. (Canadian Parks Service, Sault Ste. Marie Canal Office)

FIGURE 34. A pontoon gate lifter stepping a main lower gate leaf into position, 8 October 1894. To facilitate the gate erections, the lower entrance cotter dam was breached to flood the lock chamber. (Canadian Parks Service, Sault Ste. Marie Canal)

FIGURE 35. View of new lock after pumping out, 26 October 1894. The vertical and horizontal screw machines are being installed on the lock wall coping. (Canadian Parks Service, Canal Records, File C-4272/S32, Vol. 9)

On 24 September 1894, water was let into the lower reach of the canal and the lock chamber. The gates, built on the floor of the lock, were raised and stepped in place using a pontoon gate lifter. Shortly thereafter, the upper reach was flooded (Figs. 34 and 35), and on 15 October 1894 the steam tug *Rooth* was locked through manually. For this first lockage, the powerhouse machinery, power transmission system, and the lock gate and culvert sluice valve motors were not yet operational.[151] All of this machinery had had to be totally redesigned at the last moment in keeping with a momentous decision taken the previous summer.

To maximize the speed of operation of the new Canadian lock, the planned hydraulic system of operation had been abandoned in favour of adopting a new power source that promised even greater efficiency — electricity. To date no canal was electrically operated anywhere in the world. Nonetheless the engineers were confident that electrical power, generated on site, could be harnessed to operate the gates and sluices of the gigantic Sault Ste. Marie Canal lock.[152]

"by far the best design"

As initially designed in the winter of 1891-92, the Sault Ste. Marie Canal long lock, 900 feet by 60 feet with 19 feet of water on the sills, was unique in the world as to both its novel configuration and unprecedented length.[153] With the exception of the projected 800 foot by 100 foot Poe lock on the St. Mary's Falls Ship Canal, the Canadian lock also far outstripped the lock chamber dimensions of large ship canal locks under construction elsewhere in the world, and dwarfed the 270 foot by 45 foot locks being constructed on the Canadian ship canal system in keeping with the ongoing canal enlargement program. Indeed, the subsequent deepening of the Canadian canal to match the American Poe lock with 20 feet three inches of water on the sills, merely increased the substantial discrepancy in depth over the 14-foot draught standard of the other Canadian ship canals.

With the exception of its scale and configuration, all other aspects of the proposed Canadian long lock had been derived from traditional lock construction practice in Canada and elsewhere, or represented modifications of innovations introduced earlier on the American St. Mary's Ship Canal locks. Both the floor culverts system and the proposed hydraulic system of operation, constituted innovations in Canadian canal construction, but were of American origin. The several changes made as a result of developments in the fall of 1892 and the following summer, however, introduced unprecedented approaches to canal lock construction and operation.

The construction of lock walls with stone masonry facings and a core wall consisting of a form of rubble concrete, was unique in ship canal building. Elsewhere, plans had been prepared for constructing locks with concrete backing walls faced in brick or cut stone — brick on the English Manchester Ship Canal and stone on the Canadian Soulanges Canal — but rubble concrete core walls were unknown in lock construction.

At first glance, the Sault Ste. Marie lock walls appear to constitute an intermediate step in a natural evolution: from the traditional solid stone masonry lock wall, faced in cut stone with a rough stone backing; to the rubble concrete core wall, faced with cut stone and rough backing stone, introduced on the Sault Ste. Marie lock in 1893-94; to the concrete backing wall faced in cut stone, planned for construction on the new Soulanges Canal;[154] and ultimately to the solid concrete lock walls built on the Trent Canal, commencing in 1896, as part of an on-going effort to complete that older barge canal system through from Georgian Bay to Lake Ontario to supplement the grain-carrying capacity of the rail transport system along that route.[155] But this was not the case.

The Soulanges Canal lock plans were completed before the decision was made to introduce concrete into the walls of the Sault Ste. Marie Canal lock. It was this precedent, and system of construction, that the government engineers had in mind to speed construction when the use of concrete was sanctioned for the lock walls of the Sault Ste. Marie Canal. The novel rubble concrete core wall method of construction was simply an expedient introduced by the contractor to take advantage of a peculiar set of circumstances.

Prior to the introduction of structural concrete in the Sault Ste. Marie Canal lock, Canadian engineers had been comparatively slow in adopting concrete as a building material. With the exception of Martin Murphy, a British engineer who built a significant number of bridge piers and abutments of concrete for the Nova Scotia Department of Highways commencing in the late 1880s, in Canada concrete had been confined to use as an infill in foundation work. In effect, in situations where it was only in compression. Indeed, Canadian engineers had proved loath to use concrete in a structural capacity, or above water, through a fear of frost action on concrete, a general unfamiliarity with its behaviour, and/or a concern about the consistency of natural cement.[156]

The use of rubble concrete in the Sault Ste. Marie Canal lock walls, and subsequently plain concrete lock walls faced in stone on the Soulanges Canal, and the all-concrete walls of the Trent Canal locks, marked the beginning of the era of concrete construction in Canada. With the contemporaneous establishment of a significant number of Portland Cement plants, Canadian engineers in the first decade of the 20th century, began using plain concrete in all manner of structures — canal locks, weirs, bridge piers and abutments, concrete arch bridges, retaining walls, dock walls, and buildings of concrete block construction.[157] But it was the canal engineers, with the exception of Murphy's isolated work in Nova Scotia, who led the way in the acceptance of concrete as a building material initially challenging and ultimately superseding stone masonry.[158] The structural use of concrete in the walls of the canal and lock of the Sault Ste. Marie Canal was a milestone that helped both instigate and strengthen this trend.

As of the summer of 1894, the Sault Ste. Marie Canal lock was recognized as one of the finest masonry locks in the world, with its superiority attributed in large part to the use of concrete. And where the generation and application of electrical power was concerned, the Sault Ste. Marie canal engineers were preparing to establish a world's first, adopting a comparatively new electrical power technology to a novel purpose.

Endnotes

1   In 1887, the government supply estimates set aside $1 million for construct-
        ing a canal at Sault Ste. Marie. Construction was authorized by an order-in-
        council of 2 May 1888. (Dennis Carter-Edwards, "The Sault Ste. Marie
        Canal," *Research Bulletin*, No. 119, Environment Canada, Canadian Parks
        Service, January 1980).

2   Vernon C. Fowke, *The National Policy and the Wheat Economy* (Toronto:
        University of Toronto Press, 1957), pp. 3-4, 25-45 and 62-66; and W.T.
        Easterbrook and H.G.J. Aitken, *Canadian Economic History* (Toronto: Mac-
        millan, 1958), pp. 383-388. The immediate aims of the national policy were
        two-fold: to meet the threat of American economic penetration, and ultimate-
        ly absorption, posed to the prairie territories by the increasingly northward
        expansion of feeder railways tying into the existing American transcontinen-
        tal railway system; and to enable the new Canadian confederation to survive
        as a viable economic and political entity on the North American continent.

        The Hudson's Bay Company lands in the Northwest had been trans-
        ferred to Canada in 1870, and the Dominion Lands Act of 1871-72 provided
        for free homesteading on the prairies but reserved a 40-mile strip straddling
        the proposed transcontinental railway for railway purposes. On the national
        policy and Canadian western development see, Gerald Friesen, *The
        Canadian Prairies, A History* (Toronto: University of Toronto Press, 1984),
        pp. 162-194.

3   Fowke, as well as Easterbrook and Aitken, (op. cit.) stress the critical role
        of the transcontinental railway in the national policy program but either to-
        tally ignore, or make merely passing reference to, canal transportation.

        One of the terms of entry on British Columbia joining the Canadian con-
        federation in 1871 called for the Canadian government to undertake the con-
        struction of a transcontinental railway in two years, and build it within a
        decade.

4   Easterbrook and Aitken, *Canadian Economic History*, pp. 255, 295 and 369.

5   John P. Heisler, *The Canals of Canada* (Ottawa: Parks Canada, 1973), p.
        127.

6   Easterbrook and Aitken, *Canadian Economic History*, p. 312.

7   Allan R. Davis, C.E., "The St. Lawrence Canal Route," *Canadian Magazine*,
        Vol. 3 (1894), pp. 148-153.

8   Easterbrook and Aitken, *Canadian Economic History*, pp. 253-270. Ocean
        freight rates from Montreal were almost twice what shippers paid from New
        York. Shippers using New York were assured of balanced cargoes; whereas
        east-bound ocean ships often arrived at Montreal in ballast. Insurance rates

were also much higher on the more northerly and confined Gulf of St. Lawrence route (ibid., pp. 295 and 369).

9 Ibid., pp. 295-318; and Heisler, *The Canals of Canada*, pp. 114-116. The Grand Trunk Railroad also had two major disadvantages. It was built to a five foot six-inch gauge, necessitating transshipments of grain from the four foot 8.5-inch gauge of American railways, and freight had to be ferried across the St. Clair River at Sarnia to connect with the Detroit line.

10 Heisler, *The Canals of Canada*, pp. 120-131. The locks on the existing St. Lawrence canals system, completed in 1848, had nine feet of water on the sills but varied slightly in size: 200 foot by 55 foot locks on the Beauharnois and Cornwall canals; and 200 foot by 45 foot locks on the others. The Second Welland Canal had been reconstructed (1842-48) with locks 150 feet by 26.5 feet with 8.5 feet of water on the sills, and in 1853 had been deepened to 10 feet through raising the lock walls and embankments. The St. Lawrence canals were designed for sidewheel steamers, and the Welland Canal for the newer propeller steamers and sailing vessels (ibid., pp. 92-97, and 116).

11 "The Canadian Ship Canal Lock at Sault Ste. Marie, Ontario," *Engineering News and American Railway Journal* [hereafter cited as *Engineering News*], Vol. 33, No. 13 (28 March 1895), p. 205.

12 Heisler, *The Canals of Canada*, p. 129; and Thomas Keefer, "The Canals of Canada," *Transactions of the Royal Society of Canada*, Section III (1893), p. 29.

13 Fowke, *The National Policy*, pp. 47-48; and Easterbrook and Aitken, *Canadian Economic History*, pp. 393-394.

14 The Third Welland Canal was completed in 1883 with a 12-foot depth in keeping with the 1873 contract. In 1883-87, however, it was given a 14-foot depth through raising the lock walls and embankments. The Lachine Canal was excavated to a 12-foot depth, but equipped with locks having 14 feet of water on the sills. Between 1894 and 1899, the canal was deepened to 14 feet throughout. All canal enlargement contracts let after 1875 provided for the 14-foot depth.

15 "The Canadian Ship Canal Lock at Sault Ste. Marie, Ontario," *Engineering News* (28 March 1895), p. 205. As initially conceived, the enlarged canal system was expected to foster the growth of a substantial interprovincial trade between central Canada and the Maritime provinces, as well as a major trading system linking the inland provinces, the Maritimes, and the West Indies. During the winter months, Canadian lake freighters would work the Halifax-West Indies sea route (ibid., p. 128).

16 Environment Canada, Canadian Parks Service, Ontario Region, Canal Records [hereafter cited as Canal Records], File C-4250/S32-1, Vol. 1, John

Page, Chief Engineer, Department of Railways and Canals, to J.H. Pope, Minister of Railways and Canals, 26 April 1888.

17   Harlam Hatcher and Erich A. Walter, *A Pictorial History of the Great Lakes* (New York: Crown Publishers, 1963), pp. 263-264. The new canal superseded an American tramway constructed over the mile long portage in 1850. It had horse drawn cars running on iron strap wooden rails (ibid., pp. 261-262).

18   John H. Goff, "History of the Saint Mary's Falls Canal," in Charles Moore, ed., *The Saint Marys Falls Canal* (Detroit: 1905 Semi-Centennial Commission, 1907), p. 127; and "The New United States Government Lock at Sault Ste. Marie, Michigan," *Engineering News* (16 September 1895), p. 194.

19   Hatcher and Walter, *A Pictorial History of the Great Lakes*, p. 265.

20   Hon. Peter White, "Historical Address," in Charles Moore, ed., *The Saint Mary's Falls Canal*, Statistical Table, pp. 31-32. American pig iron production had not exceeded 600 000 tons in any year prior to 1855. With the opening of the St. Mary's Falls Canal, and the subsequent development of the Lake Superior iron mines, American pig iron production jumped to 700 159 tons in 1855, grew to over one million tons in 1864, and attained almost four million tons by 1880 (ibid.).

21   "The New United States Government Lock at Sault Ste. Marie, Michigan," *Engineering News* (16 September 1895), p. 194; and "Accident to the Sault Ste. Marie Canal Lock," *The Engineering and Mining Journal* (30 August 1890), p. 241.

22   Goff, "History of the Saint Mary's Falls Canal," p. 141. The canal was enlarged under the superintendence of General Godfrey Weitzel, U.S. Corps of Engineers. The total cost included the sum of $95 500 for a movable swing bridge wicket dam to protect the two canal lock systems (ibid.). See "The St. Mary's Falls Canal Bridge and Dam," *Engineering and Building Record and Sanitary Engineer* [hereafter cited as *Engineering Record*] (22 March 1890), pp. 246-248.

23   James P. Barry, *Ships of the Great Lakes, 300 Years of Navigation* [hereafter cited as *Ships of the Great Lakes*] (Berkeley, California: Howell-North Books, 1974), pp. 79-80, 93 and 190-126; and George A. Cuthbertson, *Freshwater: A History and a Narrative of the Great Lakes* [hereafter cited as *Freshwater*] (Toronto: Macmillan, 1931), p. 247. One upper lakes sidewheeler, the *Western World* launched at Buffalo in 1854, was 337 feet in length and reputed to be the largest steamer afloat in the world (Barry, op. cit., p. 79).

24   Barry, op. cit., pp. 128-130, and p. 136; and Karl Kuttruff, R.E. Lee, and D.T. Glick, *Ships of the Great Lakes: A Pictorial History* (Detroit: Wayne

State University Press, 1976), Figs. 13 and 14. The first iron-sheathed wooden ship on the Great Lakes was the *Merchant*, an 190-foot vessel launched at Buffalo in July 1862 (Barry, *Ships of the Great Lakes*, pp. 87-88).

25   Keefer, "The Canals of Canada," p. 41.

26   Peter White, "Historical Address," iron ore statistics, p. 32; and Brian S. Osborne and Donald Swainson, *The Sault Ste. Marie Canal, A Chapter in the History of Great Lakes Transport* (Ottawa: Environment Canada, Canadian Parks Service, 1986), p. 43.

27   R. Percy Sellon, "Electric Light Applied to Night Navigation upon the Suez Canal," *The Telegraphic Journal and Electrical Review* (14 September 1888), pp. 279-281. On the Suez Canal, freight was averaging between seven and eight million tons a year during the mid-1880s (ibid.).

28   Goff, "History of the Saint Mary's Falls Canal," pp. 167-168.

29   Ibid., pp. 167-176; and "The New United States Government Lock at Sault Ste. Marie, Michigan," *Engineering News* (16 September 1895), p. 194.

30   "The Canadian Lock at Sault Ste. Marie," *Engineering News* (20 June 1895), p. 399.

31   Canal Records, File C-4250/S32-1, Vol. 1, John Page, Chief Engineer, Department of Railways and Canals, to J.H. Pope, Minister of Railways and Canals, 26 April 1888.

32   Doug Owram, *Promise of Eden: The Canadian Expansionist Movement and the Idea of the West, 1856-1900* [herafter cited as *Promise of Eden*] (Toronto: University of Toronto Press, 1981), pp. 160-169.

33   Ibid., pp. 162-168, and 200. An "expansionist spirit," prevailing in Canada during the late 1870s and early 1880s, had led to increasingly enthusiastic estimates of the cultivable land potential of the Canadian prairies — from an estimate of 80 million acres in 1877, to 150 million acres in 1881, to as many as 300 million acres thereafter (ibid., p. 163).

34   *The Times Atlas of World History*, pp. 222-223.

35   D.A. MacGibbon, *The Canadian Grain Trade* (Toronto: MacMillan of Canada, 1932), p. 27; and Innis, *Canadian Pacific Railway*, pp. 111-112.

36   Barry, *Ships of the Great Lakes*, pp. 137-138 and 141; and Cuthbertson, *Freshwater*, pp. 268-269. The first Canadian steamer through the American St. Mary's Falls Canal appears to have been the *Collingwood*, a 188-foot sidewheeler, chartered by the Canadian government in 1857. The following year, a regular service was instituted to serve the Canadian shore of Lake Superior utilizing the steamer *Rescue*, a 121-foot twin-screw steamer built at Buffalo in 1855 (Barry, op. cit., pp. 84-86).

37   Harold A. Innis, *A History of the Canadian Pacific Railway* (Toronto: University of Toronto Press, 1923 [1971 reprint]), pp. 133-134 and 144.

38   Owram, *Promise of Eden*, p. 171.

39   Friesen, *The Canadian Prairies, A History*, p. 185. In the period 1867-99, Canada would receive only 1.5 million immigrants. In contrast, the United States received 5.5 million during the 1880s alone, and Australia received 2.5 million between 1879 and 1890 (ibid.).

40   Easterbrook and Aitken, *Canadian Economic History*, p. 395. The annual rate of population growth was only 1.61 percent in the 1870s and actually dropped to 1.13 percent during the 1880s owing to a heavy Canadian emigration to the United States. Overall, the Canadian population increased from 3.6 million in 1871 to just under 5 370 000 by 1900 — a far from encouraging growth rate (ibid., p. 395).

41   Ibid., pp. 395-396; and Friesen, *The Canadian Prairies*, p. 309. In Manitoba between 1870 and 1905, an estimated 20 percent of the settlers failed to stay on the land (ibid.)

42   Owram, *Promise of Eden*, pp. 171 and 220.

43   Carter-Edwards, "The Sault Ste. Marie Canal," p. 14; and National Archives of Canada [hereafter cited as NA], RG43, Vol. 1697, File 4860, Part II, W. Shanly, C.E., to the Minister of Railways and Canals, 13 April 1891.

44   Canal Records, File C-4250/S32-1, Vol. 1, Page to Pope, 26 April 1888; and Canada, Department of Railways and Canals [hereafter cited a DRC], *Annual Report*, 1890 and 1894, "Sault Ste. Marie Canal."

45   Barry, *Ships of the Great Lakes*, pp. 170-171; and Keefer, "The Canals of Canada," pp. 40-43. The draught limiting factor on the upper lakes was in both the St. Clair-Detroit River, the so-called "St. Clair Flats," and the St. Mary's Falls Ship Canal. At Port Colborne, it took from six to eight hours to lighten a boat at a cost of 80 cents per ton (ibid.).

        Welland Canallers varied in size. The *Bannockburn*, for example, was only 244 feet in length with a 40-foot beam, and an 85 000 bushel carrying capacity (Barry, *Ships of the Great Lakes*, p. 171).

46   Canal Records, File C-4250/S32-1, Vol. 1, Page to Pope, 26 April 1888; and NA, RG43, Vol. 1697, File 4860, Part II, W. Shanly, C.E., to the Minister of Railways and Canals, 13 April 1891.

47   Ibid., W.G. Thompson, to A.P. Bradley, 19 September 1890.

48   Canal Records, File C-4250/S32-1, Vol. 1, Page to Pope, 26 April 1888.

49   "The Canadian Canal at Sault Ste. Marie," *Engineering News* (20 June 1895), p. 399.

50   Canada. Parliament. House of Commons. *House of Commons Debates*, (Ottawa: Queen's Printer, 1888), Vol. 2, pp. 1443-1444, as cited and quoted in Carter-Edwards, "The Sault Ste. Marie Canal," p. 13.

   In the Treaty of Washington of 1871 (Article XXVII) the United States Government gave British subjects equal access to the St. Clair Flats Canal and agreed to urge the state legislatures to do likewise on their canals. In March 1872, a joint resolution of the Michigan legislatures gave British subjects equal access to the St. Mary's Falls Ship Canal.

51   Ibid., p. 13; and "The Canadian Canal at the Sault Ste. Marie," *Engineering News* (20 June 1895), p. 398. In undertaking the transcontinental railway, the Conservatives had also insisted on an all-Canadian route even though this involved considerably greater costs and difficulties in getting the railway constructed than if routed partially through American territory (Easterbrook and Aitken, *Canadian Economic History*, pp. 414-417 and 426-432).

52   Goff, "History of the Saint Mary's Falls Canal," pp. 162-163. The dispute revolved around the terms of the Treaty of Washington. The Canadian government maintained that all shipping was being equally treated in keeping with the terms of the treaty: American ships paid the same toll as Canadian ships and received the same rebate if they continued to Canadian ports (ibid.).

53   Ibid., pp. 159-161; and Carter-Edwards, "The Sault Ste. Marie Canal," pp. 9-12. Quote attributed to Sir Charles Tupper, (ibid., p. 14).

54   DRC, *Annual Report*, 1889, p. cxi; and Canals Records, File C-4250/S32-1, Vol. 4, Hugh Ryan & Co. to John Page, Chief Engineer, 30 December 1889.

55   Keefer, "The Canals of Canada," pp. 32-33. The existing St. Lawrence canals had locks of about nine-foot lift on average, and the enlarged Welland Canal had locks averaging about 13 feet of lift (ibid.).

56   W.P. Kibbee, "The Busiest Canal in the World, *Engineering Magazine* (July 1897), p. 601. Another precaution, borrowed from the Americans, was the provision for an emergency swing bridge dam to close off the canal prism in the event of serious injury to the lock gates. (See herein, Essay III - *The Emergency Swing Bridge Dam, 1895-1985*.)

57   DRC, *Annual Report*, 1889, p. cxi. The 600 foot by 85 foot lock was designed by John Page. In his capacity as Chief Engineer of the Department of Railways and Canals since 1867, Page had designed most of the locks constructed during the canal enlargement program. He was sometimes referred to as "the father of the Canadian canal system" ("The Canadian Soo and the Great Canal," *Globe*, 26 October 1895, p. 2).

58   "The Canadian Soo ...," *Globe*, 26 October 1895, p. 2.

59   Keefer, "The Canals of Canada," p. 33; and Robert W. Passfield, "Canal Lock Design and Construction: The Rideau Canal Experience, 1826-1982," (Ottawa: Environment Canada, Canadian Parks Service, Microfiche Report Series, No. 57, 1983), pp. 47-48.

60   DRC, *Annual Report*, 1889, p. cxi.

61   "The Sault Ste. Marie Ship Canal," *The Canadian Engineer* (November 1893), p. 192.

62   "The New United States Government Lock at Sault Ste. Marie, Michigan," *Engineering News* (16 September 1895), p. 194; Moore, *The Saint Mary's Falls Canal*, p. 155; and "The Sault Ste. Marie Canal," *Scientific American Supplement* (6 September 1890), p. 12231.

63   Ibid., p. 12231; and E.W. Wheeler, "Locks of the Nicaragua Canal and St. Mary's Falls Canal," *Engineering News* (1 June 1893), p. 504.

64   DRC, *Annual Report*, 1894, p. lxxvii; and "The Canadian Soo ...," *Globe*, 26 October 1895, pp. 3-4. The contracts for the upper river channel dredging and entrance piers were let to Allan & Fleming of Ottawa on 26 March 1889; and the lower river channel dredging and entrance piers to Hugh Ryan and Co. on 30 June 1889.

      On the railway construction background of Hugh Ryan, and his partners John Ryan and Michael J. Haney, see ibid., p. 4.

65   "The Sault Ste. Marie Ship Canal," *The Canadian Engineer* (November 1893), p. 191.

66   "The Canadian Ship Canal Lock at Sault Ste. Marie," *Engineering News* (28 March 1895), p. 206; and "The Sault Ste. Marie Ship Canal," *The Canadian Engineer* (November 1893), p. 191. There is no mention of the head of water obtained at the dam. Mechanical air compressors, however, worked best at low speeds, and hence under low heads of water (Joseph P. Frizell, *Water-Power, An Outline of the Development and Application of the Energy of Flowing Water*, [New York: John Wiley & Sons, 2nd. ed., 1901], p. 425).

67   "The Sault Ste. Marie Ship Canal," *The Canadian Engineer* (November 1893), p. 191; and John D. Barnett, "Pneumatic Power in Workships," ibid., Vol. 4 (July 1896), pp. 61-64.

68   Frizell, *Water Power*, pp. 411-412.

69   "The Taylor Hydraulic Air Compressor," *The Canadian Engineer*, Vol. 2 (April 1895), p. 343. A turbine powered mechanical air compressor was proposed for construction at the Newfoundland Consolidated Copper Mining Company mine at Bett's Cove, Newfoundland, as early as 1881 (*The Newfoundland Consolidated Copper Mining Company*, New York: Francis Hart & Co., 1881, p. 18).

The water-actuated Taylor air compressor, patented by C.H. Taylor of Montreal, was developed too late for use on the 'Soo' canal project, and was not really suitable for temporary installations. It was first employed at the Dominion Cotton Mills plant in Magog, Quebec, in 1896 (*The Canadian Engineer*, Vol. 4 [November 1896], "The Taylor System of Air Compression"; and ibid., Vol. 5, March 1897, "Hydraulic Air Compressor at Magog").

70 DRC, *Annual Report*, 1890, p. 162; and "The Sault Ste. Marie Ship Canal," *The Canadian Engineer* (November 1893), p. 191.

71 NA, RG43, Vol. 1698, File 4860, Part III, Hugh Ryan to Toussaint Trudeau, Chief Engineer, Department of Railways and Canals, 25 August 1892. Trudeau became Chief Engineer on the death of John Page in 1890.

72 NA, RG43, Vol. 1697, File 4860, Part II, Hugh Ryan & Co. to John Page, 23 May 1890.

73 Ibid., W.G. Thompson, Resident Government Engineer, to A.P. Bradley, Secretary, DRC, 12 August 1890.

74 Ibid., Thompson to Bradley, 7 August 1890. Allan and Fleming were also dredging the upper entrance during the summer of 1890 (Canal Records, File C-4250/S32-1, Vol. 2, Thompson to Bradley, 6 September 1890).

75 Specifications cited in ibid., W.G. Thompson to John Page, 28 November 1890; and ibid., A.P. Bradley to Hugh Ryan & Co., 10 August 1891. On the use and composition of clay puddle walls see Passfield, "Canal Lock Design and Construction: The Rideau Canal Experience, 1826-1982," pp. 108-112.

76 Canal Records, File C-4250/S32-1, Vol. 4, Hugh Ryan & Co. to John Page, 30 December 1889; and ibid., Page to Ryan, 24 January 1890. Ryan had apparently raised this question even earlier, see ibid., Thompson to Page, 28 November 1889.

77 Ibid., Ryan to Bradley, 17 October 1890.

78 NA, RG43, Vol. 1697, File 4860, Part II, W.C. Van Horne, President, CPR, to Rt. Hon. Sir John A. Macdonald, Prime Minister and Minister of Railways and Canals, 17 August 1890. See also ibid., Secretary, CPR, to Van Horne, 16 August 1890. The subtitle quote is from T. Trudeau, Chief Engineer, DRC, to Macdonald, 14 May 1891.

The contractor, Hugh Ryan, may have been the first to advocate a deepening of the canal to match the American lock. He discussed the subject with both Van Horne of the CPR and A.M. Smith of the Toronto Board of Trade during their visit(s) to the canal construction site ("Sault Ste. Marie Canal Inquiry, Minutes of Evidence," *Journals of the House of Commons*, Vol. 29, Appendix, 1895, Testimony of Hugh Ryan, 7 July 1895, pp. 135-136).

79 "The Canadian Canal at Sault Ste. Marie," *Engineering News* (20 June 1895), p. 399. Keefer maintained that the shipping capacity varied with the cube of the depth. According to his calculations, all things being equal an upper lakes freighter on a 20-foot depth of navigation would carry twice the load as on a 16-foot depth: 16 cubed versus 20 cubed (Keefer, "The Canals of Canada," p. 40).

80 NA, RG43, Vol. 1697, File 4860, Part II, W.G. Thompson, resident engineer, Sault Ste. Marie Canal project, to A.P. Bradley, 19 September 1890.

81 Desmond Morton, *A Short History of Canada* (Edmonton: Hurtig Publishers Ltd., 1983), pp. 110-112; and Peter B. Waite, *Canada 1874-1896, Arduous Destiny* (Toronto: McClelland and Stewart, 1971), pp. 221-225.

82 NA, RG43, Vol. 1697, File 4860, Part II, Thompson to Bradley, 28 March 1891; and ibid., Van Horne, Telegram, to the Rt. Hon. Sir John A. Macdonald, 30 March 1891.

83 Ibid., marginalia; "The Ship Canal," *Globe*, 13 November 1890; and ibid., "The Canadian Soo ...," 26 October 1895.

84 See NA, RG43, Vol. 1697, File 4860, Part II, W.G. Thompson, Resident Engineer, Sault Ste. Marie Canal project, to A.P. Bradley, Secretary, DRC, 16 September 1890; ibid., Thompson to Bradley, 28 March 1891; and ibid., W.G. Thompson, Reply, for the information of the Minister of Railways and Canals, to Van Horne, Telegram, 30 March 1891.

85 Ibid., Thompson to Bradley, 19 September 1890.

86 Ibid., W. Shanly, Consulting Engineer, to Minister of Railways and Canals, 13 April 1891.

87 Ibid., T. Trudeau, Chief Engineer of Canals, DRC, to the Rt. Hon. Sir John A. Macdonald, Minister of Railways and Canals, 14 May 1891.

88 DRC, *Annual Report*, 1894, p. lxxvii.

89 NA, RG43, Vol. 1698, File 4860, Part III, Trudeau to John Haggart, Minister of Railways and Canals, 29 March 1892. Macdonald had died on 6 June 1891.

The 21 feet of water on the Poe lock sills, measured by American depth standards, was only 18 feet 11 inches by the Canadian extreme low river level standard. Hence the 19 feet of water on the Canadian lock sills would be one inch more than on the American lock sills.

90 "The Canadian Soo ...," *Globe*, 26 October 1895, p. 4; and NA, RG43, Vol. 1697, File 4860, Part II, Trudeau to Hugh Ryan & Co., 29 May 1891.

91 NA, ibid., and Environment Canada, Canadian Parks Service, Sault Ste. Marie Ship Canal Office [hereafter cited as SSM Canal Office], Engineering Drawings, Drawing: "Sault Ste. Marie Canal," T. Trudeau, Deputy Minister and Chief Engineer, 28 May 1891. See also the same plan in NA, RG43,

Vol. 1697, File 4860, Part II. Doing away with the side wall culvert entrances saved 40 feet of lock wall masonry extending above the upper gates, almost the amount of extra masonry required to lengthen the lock chamber 50 feet. The canal prism remained the same width — 150 feet wide at water level.

On the floor culvert system of the American Weitzel lock see: "Accident to the Sault Ste. Marie Canal Lock," *The Engineering and Mining Journal* (30 August 1890), pp. 241-242; and E.S. Wheeler, "Locks of the Nicaragua Canal and the St. Mary's Falls Canal," *Engineering News* (1 June 1893), p. 504.

92   NA, RG43, Vol. 1697, File 4860, Part II, Thompson to Bradley, 13 January 1891, and attached drawing: "Cross-Section at right angles to Centre line of Sault Ste. Marie Canal"; and Canal Records, File C-4250/S32-1, Vol. 4, Bradley to Hugh Ryan & Co., 6 July 1891.

The retaining walls were to be nine feet three inches thick at the base and a minimum of 27 feet high, with a 2-1/2 inch-per-foot batter on the water face ("Cross-Section" drawing, op. cit.).

93   DRC, *Annual Report*, 1894, p. lxxvii.

94   "The Canadian Soo ..., *Globe*, 26 October 1895.

95   Ibid., and NA, RG43, Vol. 1698, File 4860, Part III, Trudeau to Mackenzie Bowell, Acting Minister, DRC, 17 December 1891.

96   "Steel Lock Gates for 800 x 100 Ft. Ship Canal Lock, Sault Ste. Marie, Mich.," *Engineering News* (6 August 1896), pp. 84-86.

97   NA, RG43, Vol. 1698, File 4860, Part III, Trudeau to Bowell, 17 December 1891; and ibid., Clerk of Privy Council, Report of the Committee of Privy Council to the Governor General of Canada, 24 December 1891. The cost estimates were as follows: $1 152 000 for the 650 foot by 100 foot lock with off-set gates 60 feet wide; $1 711 000 for the straight lock of that size; $1 770 000 for the 100 foot by 60 foot lock; $1 521 000 for the 830 foot by 60 foot lock; and $1 600 000 for the 900 foot by 60 foot lock.

98   Ibid., 24 December 1891; and DRC, *Annual Report*, 1894, p. lxxvii.

99   NA, RG43, Vol. 1698, File 4860, Part III, F.H. Langer, Buffalo, to Frank Kirby, Detroit, 11 November 1891.

100   "The Canadian Ship Canal Lock at Sault Ste. Marie, Ont.," *Engineering News* (28 March 1895), p. 205. See also Keefer, "The Canals of Canada," p. 31.

101   Keefer, op. cit., pp. 28-29 and 40.

102   Cuthbertson, *Freshwater*, p. 269. The three original CPR passenger steamers had been constructed in Scotland and designed as ocean steamers, a radical departure for Great Lakes ships. They were cut in half to pass through the

St. Lawrence and Welland Canals, and then reassembled at Buffalo. In 1883-84, there were no large Canadian shipyards on the upper lakes (ibid., p. 268).

103 NA, RG43, Vol. 1698, File 4860, Part III, Trudeau to Bowell, 17 December 1891; and Keefer, "The Canals of Canada," p. 41. At least two steam barges exceeded 320 feet by 43 feet: the 300 foot by 45 foot *Western Reserve* launched in 1890, (Barry, *Ships of Great Lakes*, p. 145); and the 456 foot by 50 foot *D. Houghton* built in 1889 (Cuthbertson, *Freshwater*, p. 264). With the exception of these two vessels, the overall trend as of 1891 had not been toward increasing breadth of beam.

104 Barry, *Ships of the Great Lakes*, pp. 153-155.

105 NA, RG43, Vol. 1698, File 4860, Part III, Trudeau to John Haggart, Minister of Railways & Canals, 29 March 1892; and DRC, *Annual Report*, 1893, p. 121.

106 "The Sault Ste. Marie Ship Canal," *The Canadian Engineer* (November 1893), p. 192; and SSM Canal Office, Drawing No. 490-1, Case D - 155, "Sault Ste. Marie Canal, Plan of Lock as Enlarged," T. Trudeau, Chief Engineer of Canals, 5 April 1892. See also Keefer, "The Canals of Canada," Plate IV.

107 NA, RG43, Vol. 1697, File 4860, Part II, Hugh Ryan & Co. to A.P. Bradley, 23 May 1890; ibid., W.G. Thompson to Hugh Ryan & Co., 9 July 1890; and ibid., Vol. 1698, File 4860, Part III, "A - Specifications for Lock Backing," 5 April 1892.

In "pinning" smaller slender stones, or quarry chips, were driven into the vertical and horizontal joints of rough masonry walls to wedge the stonework into a solid mass.

108 Ibid., Part V, Report of Privy Council, 27 August 1895.

109 Christopher C. Stanley, *Highlights in the History of Concrete* (London: Cement & Concrete Association, 1979), pp. 9-10.

110 Carl W. Condit, *American Building, Materials and Techniques from the Colonial Settlements to the Present* (Chicago: University of Chicago Press), 1968, pp. 155-157; and Norman Davey, *A History of Building Materials* (London: Phoenix House, 1961), pp. 97-98 and 102-105.

111 Davey, op. cit., pp. 106-108; and Stanley, *Highlights in the History of Concrete*, pp. 11-17.

Hydraulic cement was first used in canal construction for the masonry locks of the Calder and Hebble Navigation in 1760-64. They were built by John Smeaton following his rediscovery of the properties of hydraulic cement, and its employment in building his renown Eddystone Lighthouse in 1758-59 (ibid.; and Passfield, "Canal Lock Design and Construction ...," p. 121).

112   "Sault Ste. Marie Canal Inquiry," op. cit., Hugh Ryan testimony, p. 109.

113   Passfield, "Canal Lock Design and Construction ...," pp. 122-129. Initially, the Rideau Canal lock masonry was laid in common lime mortar, manufactured on site, and pointed with imported English natural cement. This proved unsatisfactory as the imported natural cement tended to spoil before reaching the work sites (ibid.).

114   Thomas Ritchie, *Canada Builds, 1867-1967* (Toronto: University of Toronto Press, 1967), pp. 231 and 233.

115   NA, RG43, Vol. 1698, File 4860, Part III, Trudeau to Hugh Ryan & Co., 5 April 1892; and "The Canadian Ship Canal Lock at Sault Ste. Marie," *Engineering News* (28 March 1895), p. 206.

116   "The Canadian Ship Canal Lock ...," op. cit.; and SSM Canal Office, Drawing No. 490-1, Case D-155, "Sault Ste. Marie Canal, Plan of Lock as Enlarged," T. Trudeau, Chief Engineer, 5 April 1892. See also Canal Records, File C-4272/S32, Vol. 9, Photo of floor culverts under construction, ca. 1894.

117   See "The New United States Government Lock at Sault Ste. Marie, Mich.," Part I, *Engineering News* (26 September 1895), pp. 194-196; ibid., Part II (10 October 1895), pp. 238-239; and "Accident to the Sault Ste. Marie Canal Lock," *The Engineering and Mining Journal* (30 August 1890), pp. 241-242.

118   SSM Canal Office, Drawing: "Sault Ste. Marie Canal, Proposed Lock Machinery," Longitudinal Section and Elevation at Upper Gates, 15 January 1891; Canal Records, File C-4272/S32, Vol. 2, "Sault Ste. Marie Canal, Part Details of Lock Showing Culverts etc. for Valve Shafts," T. Trudeau, 23 August 1892; and NA, RG43, B2e, Vol. 1771, Fol. 332, DRC "Specification for Valves, including frames, shafts, pillow blocks, crank arms, draw rods, securing bolts, grating, etc.," 30 November 1893.

119   In the trussed bow-string gate, the horizontal rails were framed into the quoin and mitre posts in the conventional manner with mortise and tenon joints, but each rail was a framed truss built up of an arched bar chord and straight bar chord framed around three vertical intermediate posts and trussed together with iron rods.

      On the lock gates see: NA, RG43, B2e, Vol. 1771, Fol. 319, DRC, "Specification of Work to be done in the construction of the Lock Gates," 10 December 1894; and SSM Canal Office, Engineering Drawings, Case N, Drawing No. 54547, "Sault Ste. Marie Canal, Plan of Upper Main Spare Gates," 10 December 1894.

120   "Steel Lock Gates for 800 x 100 Ft. Ship Canal Lock, Sault Ste. Marie, Mich.," *Engineering News* (6 August 1896), p. 84; and "The Canadian Ship Canal Lock at the Sault Ste. Marie," *Engineering News*, 28 March 1895, p. 208.

121 NA, RG43, Vol. 1698, File 4860, Part III, Trudeau to John Haggart, Minister of Railways and Canals, 29 March 1892; and ibid., Trudeau to Hugh Ryan & Co., 5 April 1892.

122 J.B. Spence, "Power Transmission, The Canadian Ship Canal Lock at Sault Ste. Marie and Its Electrical Operation," [hereafter cited as "Power Transmission, *Electrical Engineer*], *The Electrical Engineer, A Weekly Review of Theoretical and Applied Electricity* (16 October 1895), p. 380. Although the gates on the 'Soo' canals were extremely large by contemporary standards, in both Britain and Europe by the early 1890s several coastal harbour tidal locks were being constructed with timber, iron, and steel gates of various types that approached the Poe lock in width and exceeded it in depth ("Steel Lock Gates ...," *Engineering News* (6 August 1896), p. 86, "Table Showing Dimensions, Cost, Etc. of St. Mary's Falls Lock Gates as Compared with Other Large Lock Gates in America and Europe").

123 "The Canadian Ship Canal Lock at Sault Ste. Marie, Ont.," *Engineering News* (28 March 1895), p. 207 and Fig. 6, "Plan Showing Attachment and Arrangement of Cables Operating Gate Leaves."

124 "Accident to the Sault Ste. Marie Canal Lock," *The Engineering and Mining Journal* (30 August 1890), p. 241, Fig. 5, "Plan of Gates and Machinery." Evidence of the decision to adopt the American hydraulic gate machines can be seen in the 5 April 1892 plan of the new 900 foot by 60 foot lock chamber.

125 DRC, *Annual Report*, 1893, pp. 121-122.

126 Goff, "History of the Saint Mary's Falls Canal," pp. 163-164; and DRC, *Annual Report*, 1893, pp. 286-289. The American tolls were imposed as of 1 September 1892.

127 "The Sault Ste. Marie Ship Canal," *The Canadian Engineer*, Vol. 1 (November 1893), p. 191; and "The Canadian Soo ...," *Globe*, 26 October 1895, p. 4.

128 NA, RG43, Vol. 1698, File 4860, Part III, Hugh Ryan & Co. to Trudeau, 9 September 1892; ibid., Trudeau to John Haggart, Minister of Railways and Canals, 14 October 1892. The subtitle is from the terms of the 8 November 1892 agreement quoted in Canada, *Journals of House of Commons, Appendix*, Vol. 29, 1895, "Sault Ste. Marie Canal Inquiry, Minutes of Evidence," [hereafter cited as "Sault Ste. Marie Canal Inquiry"], 17 May 1895, p. 9.

129 DRC, *Annual Report*, 1892, pp. 131-132. Thomas Munro was the engineer responsible for designing the Soulanges Canal works. It was Munro who prepared a report on the Manchester Ship Canal, and advocated the extensive use of concrete in the new Soulange Canal.

On the Manchester Ship Canal, there were five pairs of locks — the largest a 600 foot by 80 foot tidal lock — under construction from 1887 to 1894 to enable ocean vessels to proceed 35 miles inland on a 26 feet deep navigation. The locks were of solid concrete construction, faced with blue brick and a fender course of granite above water level. The sills, hollow quoins, and the roller wheel track quadrant below each gate leaf were also in granite. The gates and sluice valves were hydraulically operated ("The Manchester Ship Canal," *Engineering, An Illustrated Weekly Journal* [London], 25 April 1890, p. 497, 30 May 1890, p. 640, and 6 June 1890, pp. 682-684.)

130 NA, RG43, Vol. 1698, File 4860, Part III, Thompson to Secretary, DRC, 9 September 1892. The government proposed the use of concrete in response to Hugh Ryan's initial assertion that the stone masonry lock walls could not possible be completed as of 1 December 1893 (ibid.).

131 "The Canadian Soo ...," *Globe*, 26 October 1895, p. 3.

132 "The Sault Ste. Marie Ship Canal," *The Canadian Engineer* (November 1893), p. 191; and "Sault Ste. Marie Canal Inquiry," 25 June 1895, p. 107. Of the 40 derricks, 33 were steam-powered and the rest presumably were worked by the compressed air system. In addition, there were about a dozen horse-powered hoists ("The Canadian Soo ...," *Globe*, 26 October 1895).

133 "The Sault Ste. Marie Ship Canal," *The Canadian Engineer* (November 1893), p. 191. Augmenting the workforce caused wage rates to soar. The various trades took advantage of the heavy labour demand to strike for higher wages in the spring of 1893. Wages, which had been 25 percent above the general rate for each trade, were pushed up as much as 40 percent more and in one case 60 percent over the going rate elsewhere. For a breakdown of the wage rates by trade, both before and after the strikes, see "Sault Ste. Marie Canal Inquiry," Hugh Ryan testimony, pp. 105-107 and 116.

134 "Sault Ste. Marie Canal Inquiry," p. 120. Ryan & Co. had intended to construct an electric plant for night work — probably through using the belting and shafting on the turbines in the existing powerhouse to drive electric generators powering arc lamps. But this was not done. The bonfires cast enough light (ibid., p. 107).

135 Ibid., p. 31 and 149-151. About 7000 cu. yards of stone masonry had been laid in the north wall prior to the spring of 1893 ("The Canadian Soo ...," *Globe*, 26 October 1895, p. 4).

136 "Sault Ste. Marie Canal Inquiry," pp. 149-150, 169-171, and 178. The stone facing of the backing wall was apparently laid without headers (ibid., p. 169).

On the supposed advantages of rubble concrete over either solid concrete or cut stone masonry wall, see "The Merits of Rubble Concrete," *Engineering News*, Vol. 50, No. 3 (16 July 1903), pp. 58-59.

137 Ibid., pp. 116, 150, and 153-155.

138 Ibid., p. 179; and "The Canadian Soo ...," *Globe*, 26 October 1895, p. 3. Apparently there were no Canadian Portland cement manufacturers when construction commenced on the Sault Ste. Marie Canal project. In 1891,
however, several plants were established in Owen Sound (Thomas Ritchie,
Canada Builds, 1867-1967, p. 231).

139 "Sault Ste. Marie Canal Inquiry," 27 June 1895, p. 116; and "The Canadian
Soo ...," *Globe*, 26 October 1895. See also NA, RG43, File 4860, Part V,
John J. McGee, Clerk, Report of Privy Council, 27 August 1895.

140 "Sault Ste. Marie Canal Inquiry," 27 June 1895, pp. 116-117; and Canal
Records, File C-4572/S32, Vol. 9, Historic photo of timber culverts under
construction, n.d. [1894].

On the 'Soo' project, almost 90 000 barrels of cement were used: 50 063
barrels of which were Portland cement used for making concrete (22 000
barrels) and mortar (28 000 barrels). The Portland cement came in 350-
pound barrels, with one barrel making approximately a cubic yard of concrete. The Canadian natural cement came in 240-pound barrels ("Sault Ste.
Marie Canal Inquiry," 27 June 1895, pp. 116-117).

In the 900 foot by 60 foot lock, there were 42 700 cu. yards of backing
wall between the piers, and 22 300 cu. yards of face stone (NA, RG43, Vol.
1698, Part III, File 4860, Part III, Thompson to Secretary, DRC, 9 September 1892).

141 In the United States, the first use of structural concrete for constructing canal
locks was on the Illinois & Mississippi (or Hennepin) Canal. This seven-
foot-deep navigation was constructed in 1892-1907 with locks of solid mass
concrete made from imported German Portland cement. Each 170 foot by
35 foot lock comprised about 3535 cu. yards of concrete. (Mary Yeater, "The
Hennepin Canal," *American Canals, Bulletin of the American Canal Society*,
No. 22 [August 1977], p. 7; and "The Illinois & Mississippi Canal Lock
Works," *Engineering News* [14 February 1895], pp. 98-101.)

142 "The Canadian Soo ...," *Globe*, 26 October 1895, p. 4.

Initially, the cut stone facing was estimated at $11.94 per cu. yard, the
backing stone at $11.60 per cu. yard, and the concrete fill in the lock foun-
dation at $7. per cu. yard. The proposed solid, plain-concrete backing wall
was estimated at $8.30 per cu. yard. Escalating wage rates, however, forced
adjustments. Ryan was paid prices of $11., $12.60, and $16. per cu. yard for
parts of his hybrid wall backing (NA, RG43, Vol. 1698, File 4860, Part III,

Thompson to Secretary, DRC, 9 September 1892; and "Sault Ste. Marie Canal Inquiry," pp. 42 and 150).

143 "Sault Ste. Marie Canal Inquiry," pp. 150-151, 156, 169, 171, and 176.

144 Ibid., pp. 56-57. Even before its completion, the natural cement in the Poe lock expanded and cracked (ibid.). Masonry work had commenced on the Poe lock in September 1891, and when completed would comprise 80 876 cu. yards of masonry ("The New United States Lock at Sault Ste. Marie, Mich.," *Engineering News* (16 September 1895), p. 195.

145 "Sault Ste. Marie Canal Inquiry," 25 and 27 August 1895, pp. 175 and 176; and DRC, *Annual Report*, 1893, p. 46.

146 NA, RG43, Vol. 1698, File 4860, Part IV, Collingwood Schreiber, Chief Engineer, DRC, to John Haggart, Minister of Railways and Canals, 10 February 1893; and ibid., Engineering Drawing, Cross-section of cribwork and masonry retaining wall, 14 July 1893.

147 "Sault Ste. Marie Canal Inquiry," pp. 43, 83, 86-88, and 125. As constructed the cribwork apparently extended almost 2600 feet along each side of the canal prism. This is almost the entire length of the canal, exclusive of the lock (ibid., p. 86).

148 "The Canadian Soo ...," *Globe*, 26 October 1895, p. 3; and "Sault Ste. Marie Canal Inquiry," pp. 43, 83 and 125.

The concrete retaining wall expedient was afterwards regarded as a significant improvement. It would protect the fractured sandstone rock face of the canal excavation from erosion below the cribwork (DRC, *Annual Report*, 1895, p. 130).

149 Ibid., pp. 42-43, 86-87, 110-114, and 165. The cribwork specifications are in ibid., p. 153. The masonry retaining wall was two feet six inches wide at the coping, with the 2.5-inches to one-foot batter carried down five feet on its rear face and then plumb (ibid., p. 87).

150 DRC, *Annual Report*, 1893, p. 46; and ibid., 1894, Appendix No. 6, pp. 121-122. The firms are listed, according to their contributions, in ibid.; and "The Canadian Soo ...," *Globe*, 26 October 1895, p. 3.

151 DRC, *Annual Report*, 1895, Appendix No. 7, p. 129 and p. lxxviii. In addition, a good deal of dredging remained in the approaches to the canal, as well as rock excavation at the site of the coffer dams, to deepen the canal to a 20-foot navigation depth.

152 Ibid.; and J.B. Spence, "Power Transmission, *Electrical Engineer*, 16 October 1895, pp. 380-381.

153 The subtitle is from a statement attributed to E.S. Wheeler, one of engineers on the American Poe lock construction project, in speaking of the new

Canadian lock chamber ("Sault Ste. Marie Canal Inquiry," 7 July 1895, p. 136).

154   The 14-mile-long Soulanges Canal had five locks, 270 feet by 45 feet with over 14 feet of water on the sills. It was to have been completed by 31 October 1894 but construction difficulties — a tenacious blue clay filled with huge boulders, landslides, and defaulting contractors — delayed the opening until October 1899. The first lock, of concrete faced with cut stone, was completed in June 1898 (DRC, *Annual Report*, 1898, p. 123). See also C.R. Coutlee, "The Soulanges Canal Works, Canada," *Engineering News* (18 April 1901), pp. 274-278.

155   In the fall of 1896, plain concrete walls were poured for lock no. 4 on the Peterborough-Lakefield Division of the Trent Canal. This lock — now designated lock no. 23, Otonabee — was the first Canadian lock constructed wholly in concrete (DRC, *Annual Report*, 1897, "Trent Canal," p. 140). The Trent lock predates any lock on the new Soulanges Canal.

156   M. Murphy, "Concrete as a Substitute for Masonry in Bridge Work," *Transactions, Canadian Society of Civil Engineers* [hereafter cited as Transactions, CSCE], Vol. 2, 23 February 1888, pp. 79-93, and Discussion, pp. 94-111; and DRC, *Annual Report*, 1897, "Trent Canal," p. 140.

157   Ritchie, *Canada Builds*, pp. 234-235. See also C.R. Young, "Bridge Building," *The Engineering Journal*, Vol. 20, No. 6 (June 1937), pp. 478 and 490-497; and Ann Gillespie, "Early Development of the Artistic Concrete Block: The Case of the Boyd Brothers," *APT Bulletin*, Vol. 11, No. 2 (1979), pp. 30-52.

158   On the Soulanges Canal, concrete replaced timber in the lock floors and culverts as well as the backing stone in the lock walls. It was also used extensively in constructing weirs, bridge piers and abutments, cut off walls, and retaining walls which were given merely a coping of stone (DRC, *Annual Report*, 1892, pp. 131-132; and "The Soulanges Canal Works, Canada," pp. 274-278).

Ironically, in England and the United States, canal engineers were among the last to employ structural concrete (Condit, *American Building, Materials and Techniques*, p. 159; and Stanley, *Highlights in the History of Concrete*, pp. 19-28). But Canadian canal engineers immediately picked up on the English, and American, introduction of concrete into canal construction and in the 1890s were in the forefront of concrete construction work in Canada.

After the turn of the century, the plain concrete used by the canal engineers gave way to reinforced concrete construction. The first Canadian reinforced concrete arch bridges were built ca. 1906; and by 1920 reinforced concrete was widely used for constructing bridges, grain elevators, roads, buildings, harbour works, and power dams (Frank Barber, "Canadian Reinforced Concrete Arch Bridges," *The Canadian Engineer* [13 March 1919], pp. 289-293; and Edwin Tomlin, "The Place of Concrete in the Coming Era, On the North American Continent," *Concrete & Constructional Engineering* [1920], pp. 20-22 and 101-102).

II. ELECTRIFICATION AND OPERATION, 1893-1985

Introduction

In 1893, as work proceeded in constructing the novel 900 foot by 60 foot long lock and powerhouse on the Canadian Sault Ste. Marie Ship Canal, doubts arose concerning the efficiency of the planned hydraulic system of operation. But a comparatively new power source — electricity — appeared to offer a solution. An experiment on another Canadian canal had proved the feasibility of employing electrical power for operating canal locks, and converting the new lock to electrical operation posed no overwhelming difficulties. As a result, the world's first electrically powered lock was placed in operation in September 1895, ushering in a new era of ship canal construction and operation.

The electrically powered long lock proved highly efficient, and for a time served as a major link on by far the world's busiest canal system. Despite employing an electrical technology rapidly approaching obsolescence, the original power system continued to perform flawlessly for almost half a century until modernized in response to a wartime upsurge in shipping.

Electrification

From the commencement of work on the Sault Ste. Marie Canal in the spring of 1889, plans called for the construction of an hydraulically powered lock. It was to be similar in design to the Weitzel lock in operation since 1881 on the American St. Mary's Falls Ship Canal at Sault Ste. Marie, Michigan. In the hydraulic system designed for the new Canadian lock, the powerhouse was located adjacent to the lower gates of the lock chamber. A penstock, six feet eight inches in diameter, carried water from the upper reach of the canal to the powerhouse where the 18-foot head of water would operate two vertical shaft turbines. The turbines, positioned in a single open draft box, were to be geared to two force pumps and two centrifugal pumps (Fig. 36).

The force pumps were for pumping water under pressure into an accumulator where the water in the piping system would be maintained under a constant pressure of up to 120 p.s.i. until needed to work the pistons of the hydraulic cylinders operating the gates and sluice culvert valves on the lock chamber. The centrifugal pumps, deep in the basement of the powerhouse, were to be used for dewatering the lock chamber when repairs were required.

The hydraulic system was self-contained, requiring no outside power source or costly fuel. Based on the American experience, hydraulic power had appeared the most efficient and economical means of operating a huge ship canal lock.[1] But as construction proceeded at Sault Ste. Marie, a new system was investigated.

In preparing plans for the construction of another ship canal — the Soulanges Canal intended to replace the old Beauharnois Canal bypassing the Cascades

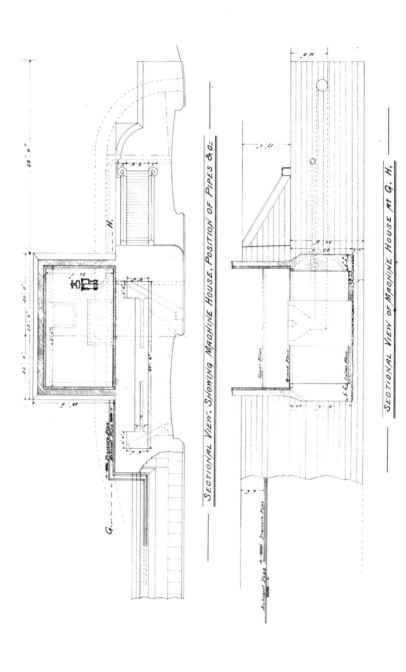

FIGURE 36. The original powerhouse layout, for the discarded hydraulic system of operation. (Detail from "Plan of Lock as Enlarged," 28 May 1891, Canadian Parks Service, Sault Ste. Marie Canal Office)

Rapids on the St. Lawrence River — engineers of the Department of Railways and Canals had begun experimenting in 1893 with using electricity as a power source.[2] At that time, no lock anywhere in the world was being operated electrically.[3]

The experiments were carried out on a gate leaf of lock no. 4 of the Beauharnois Canal using an electrical power system designed by Thomas Munro, the engineer in charge of the Soulanges Canal construction project.[4] As recently as July 1891, a small hydro-electric plant had been constructed near lock no. 4 to operate an Edison direct-current electric lighting system along the lower section of the Beauharnois Canal.[5] Munro, with the assistance of the Canadian General Electric Company, used the existing turbine to power a 45-horsepower dynamo which was connected in turn to an old street car motor of about 15 horsepower. The motor, through a heavy gear reduction system, was made to work a gate leaf by means of a pinion gear meshing in a rack on a draw/push bar consisting of a steel I-beam. The outer end of the beam was attached directly to the gate leaf near the mitre post: driving the motor in one direction drew back the beam drawing the gate open; and reversing the motor ran out the beam pushing the gate closed.

The experiments proved that little power would be required to work lock gates. The only difficulty encountered was overcoming friction in the complex gearing system that reduced the 900 r.p.m. speed of the motor down to the 10 to 15 feet per minute required on the I-beam acting as a draw/push bar. But Munro was convinced that a properly designed system, using motors adapted for the work, would easily overcome this difficulty. Moreover, electrical power would greatly speed the operation of a lock, increase safety, and reduce costs far below the manually operated system in use on Canadian canals.

As soon as the trials concluded during the summer of 1893, Munro had begun preparing plans for operating the new Soulanges Canal completely with electricity. This included the lock gates and the sluice valves of the locks as well as the regulating weirs and swing bridges. A powerhouse was to be built on the summit level where the rivière à la Graisse crossed under the canal and fell 20 feet into the St. Lawrence River.[6] Once the feasibility of adopting electrical power was proven, the canal engineers at Sault Ste. Marie began to examine the new power source as a means of overcoming a serious deficiency noted in the hydraulic system in operation on the American lock at Sault Ste. Marie.

For the better part of the navigation season, the hydraulic system worked exceptionally well. The hydraulic gates of the Weitzel lock could be closed in 1.5 minutes, and the large 515 foot by 85 foot lock chamber filled, by means of the hydraulically operated floor culvert sluice system, in only 12 minutes or emptied in 7.5 minutes.[7] In cold weather, however, this was not the case. To avoid freezing problems, the water in the gate operating cylinders on the lock walls was replaced with oil prior to the onset of freezing temperatures. But the oil thickened in cold weather making the operation of the gates very sluggish, slowing canal traffic to a crawl through the lock.[8] This was a matter of acute concern to the Canadian engineers.

In the grain trade, enormous traffic was generated at either end of the navigation season. In the fall, ships rushed to carry the fall grain crop east before the winter freeze up blocked navigation channels solid with ice. All shipping available on the lakes was then diverted to carrying grain, including even the huge iron ore carriers. Similarly in the spring shippers again strove on the opening of navigation to quickly empty the grain elevators holding the last of the previous year's crop.[9] Therefore, it was essential that the Canadian lock at Sault Ste. Marie operate at maximum efficiency during cold weather. An electrically powered lock would do this, regardless of freezing temperatures, and such a system promised to be as economical as an hydraulically powered lock.[10]

During the summer of 1893 with masonry work well advanced on the lock chamber, it was decided to convert the Sault Ste. Marie Canal to electrical operation. New machinery, powered by electric motors, would be required for working the lock gates as well as for the butterfly valves previously designed for the filling and discharge sluice culverts in the floor of the lock. This, however, involved no change in the lock gate specifications or the butterfly valves. Consequently, the contract for five pairs of lock gates was awarded as early as 13 December 1893 to the general contractor, Hugh Ryan Co. who then subcontracted the work to Justus and Roger Miller, lock gate contractors from Ingersoll, Ontario. The eight steel butterfly valves were contracted out on 8 February 1894 to Ryan, with the work being done by the St. Lawrence Foundry Co. and the Bertram Engine Works Co. of Toronto.

The Powerhouse

In the powerhouse, all new machinery had to be designed for converting the interior into an electrical power generating station. This involved changing the layout and capacity of the turbines, modifying the penstock configuration, and substituting generators or dynamos for the force pumps and accumulator of the abortive hydraulic system. The centrifugal pumps for dewatering the lock chamber, however, were retained. The contract for the modified penstock was let to Hugh Ryan & Co. on 19 December 1893; and the two dewatering pumps were contracted to Beatty & Sons of Welland, Ontario on 29 January 1894.[11]

Design work on the new electrical power plant continued during the winter of 1893-94 under the direction of James B. Spence, Chief Draughtsman of the Department of Railways and Canals.[12] The new specifications for the turbines, as well as the shafting, pulleys and belts, and gearing for the powerhouse machinery were completed and the work contracted out to William Kennedy & Sons of Owen Sound, Ontario, as of 11 January 1894.[13] On 9 May 1894, the contract for the electrical plant was let to the Canadian General Electric Company of Toronto and Peterborough, Ontario. This contract also included the installation of an electric lighting system for the canal and powerhouse. All of the contracts specified the work be completed by 1 July 1894 in keeping with the Conservative government's desire to have the canal in operation by that date.[14] This schedule, however, was not met. Delays were encountered owing to changes made in the design details of the power plant machinery after the contracts were let.[15]

On the first floor of the powerhouse, two 45-inch turbines of 155 horsepower each were installed to drive the generators and dewatering pumps. The turbines were of the "New American" type, mounted on horizontal steel shafts. To enable the turbines to run independent of each other, the six foot eight inch steel penstock was branched into two five-foot diameter pipes a short distance outside of the powerhouse. The draft tubes and discharge pipes were also five feet in diameter and discharged separately back into the canal below the lock chamber. Valves on each penstock branch, and on each draft tube, enabled the turbines to be laid dry for repairs (Figs. 33 and 37).[16]

The new, or altered, turbine and penstock layout was typical of contemporary hydro-electric generating stations. Since the mid-19th century, the traditional vertical shaft turbine had been in widespread use in North America for driving line shafting powering machinery in mills and factories.[17] A dramatic change had occurred, however, during the 1880s in response to the new electric lighting and power industry. The early generators or dynamos were invariably manufactured with horizontal shafts. Consequently horizontal shaft turbines, whether direct-connected or more generally belt-connected, lent themselves more readily to driving dynamos.

The horizontal shaft turbine setting eliminated the bevel-gearing power takeoff required to connect a vertical shaft turbine with horizontal line shafting, and also dispensed with the step bearing in which the vertical shaft turbine had rested and turned. This proved advantageous in several respects: the heavy power takeoff gearing had comprised a major component of turbine installation costs; the bevel gear friction losses had ranged from 10 to 20 percent of turbine power output; and the step bearing, generally constructed of a tallow-soaked hardwood, had proved troublesome with the on-going increases in turbine size and weight.

It was also far easier, with belt-connected horizontal shafts, to overcome the critical problem of matching the turbine speed to the r.p.m. required for operating a generator.[18] As a consequence, the traditional vertical shaft turbine had rapidly

FIGURE 37. Twin steel penstocks branching off from the main penstock to supply the two horizontal shaft turbines in the powerhouse, spring of 1894. (Canadian Parks Service, Sault Ste. Marie Canal Office)

been superseded in all applications, and by the mid-1890s three-quarters of the turbines in North America were on horizontal shafts.[19]

In this period, multiple turbines mounted on horizontal shafts were often supplied with water from a common penstock, branched to each runner. This was not the best arrangement as it caused whirling and eddying at the entrance to each turbine. But this layout was acceptable to contemporary engineers. And on the 'Soo' canal, it minimized costly changes needed to adopt the single penstock of the abortive hydraulic power plant to the new horizontal shaft turbine layout of the hydroelectric plant.

In one respect, the turbine layout was dramatically improved. The vertical shaft turbines of the hydraulic plant were to have discharged into a common draft tube. This arrangement invariably caused serious interference, reducing the efficiency of both turbines. Providing a separate draft tube and discharge pipe for each turbine overcame this deficiency (Figs. 38 and 40).[20] The comparatively large diameter of each draft tube also contributed to maintaining the efficiency of the turbine when mounted on a horizontal shaft.[21]

The "New American" turbines selected for the powerhouse installation were, as indicated by the manufacturer's name, a peculiarly American type of water

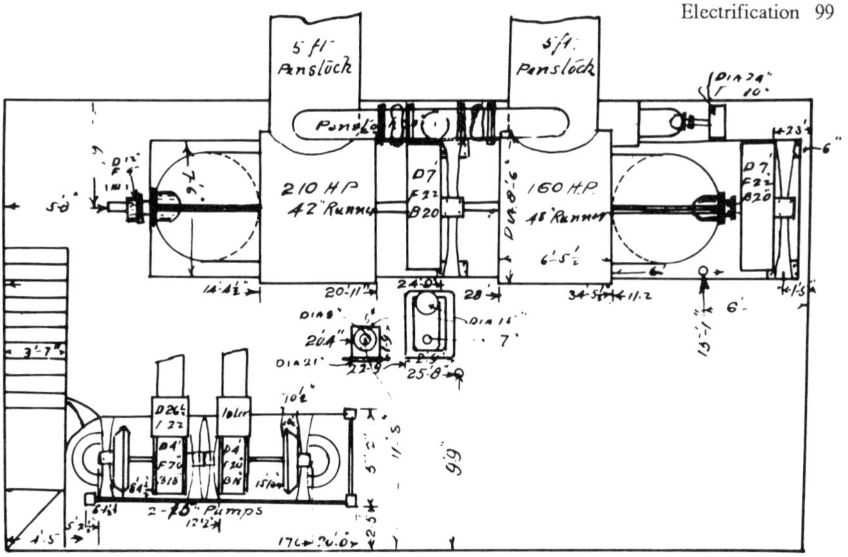

FIGURE 38. "Power House, Lower Floor," showing layout of the two main turbines and the gearing system for the centrifugal pumps below in the basement. The 210-horsepower turbine, and adjacent governor, replaced the original south turbine in 1910-11. (*Ross Notebook*, Sault Ste. Marie Canal Office, ca. 1930)

wheel. First introduced in 1859 as a major modification of the Francis inward-flow turbine and modified thereafter through a quarter century of "cutting and trying," the American-type turbine had evolved into a distinctive water wheel. In the ongoing process of adaptation and improvement, the buckets of the inward-flow Francis turbine were lengthened providing axial flow, and then curved outwards to discharge the water, and the number of buckets was substantially reduced increasing the size of the openings. The result was the unique mixed-flow American type of turbine combining the flow characteristics of three distinct types of earlier turbines: the inward-flow of the Francis, or Howd-Francis wheel; the axial or parallel flow of the Geyelin-Jonval wheel; and the outward flow of the Boyden-Fourneyron wheel. In effect, the elongated buckets of the runner of the American-type turbine were shaped so that the water flowed inwards, both laterally and axially, with a reverse curve at the discharge end that threw the water outwards again (Fig. 39).

By the 1890s, the American mixed-flow turbine — marketed under various names by a variety of manufacturers — had almost totally superseded other types of turbines for industrial and electrical power generation applications in the United States.[22] Its performance was second to none. When employed in conjunction with a large draft tube, the mixed-flow turbine had attained a highly impressive 90 percent efficiency rating whether operated in a vertical or horizontal shaft arrangement.[23] percent efficiency rating whether operated in a vertical or horizontal shaft arrangement.[23]

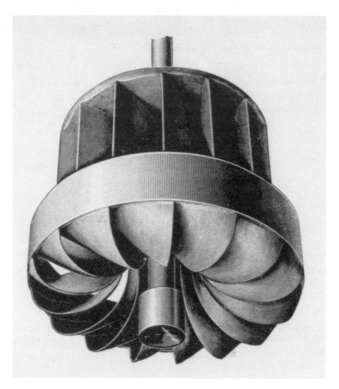

FIGURE 39. Runner of a "New American" turbine, showing the configuration of the buckets. (*Journal of the Western Society of Engineers*, April 1898)

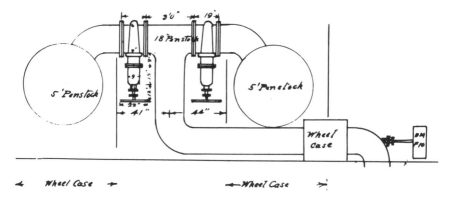

FIGURE 40. Arrangement of the 18-inch by-pass penstock, enabling the 13-inch auxiliary turbine to drive the incandescent lighting system independent of the two main turbines. A six-kW dynamo replaced the original three-kW incandescent lighting dynamo in 1921. (Ross Notebook, Sault Ste. Marie Canal Office, ca. 1930)

In Canada, the William Kennedy and Sons Company of Owen Sound, Ontario, were the sole agents and manufacturers of the "New American" version of the mixed-flow turbine. Various brand-name versions of the American mixed-flow turbine were manufactured in Canada, but this type of turbine did not predominate among water power installations.[24]

When work commenced on the Sault Ste. Marie Canal powerhouse the two 155-horsepower turbines, worked by an 18-foot head of water, constituted a major installation in comparison to the scale of existing contemporary hydro-electric power stations. As of the early 1890s, few of the stock turbines in North American power stations exceeded several hundred horsepower.[25] The very largest had a rating of just over 500 horsepower.[26]

At the 'Soo', each of the 45-inch turbines was belt connected to a horizontal countershaft up above on the second floor of the powerhouse. The horizontal countershaft or line shaft was connected in turn with oak-tanned, two-ply leather belts to three separate generators or dynamos on the second floor. The dynamos were part of two distinct direct current (d.c.) electrical generating systems: a power system and an arc lighting system (Fig. 41).

Power for the motors of the lock gate, and sluice valve, operating machinery was provided by a 45-kilowatt, 500-volt Edison bi-polar generator. A second identical generator was installed as a back up (Figs. 42 and 54). One turbine drove the main 45-kW generator. The other turbine could also be used to operate either of the electrical power generators in an emergency, but was intended to drive an arc lighting generator. This was a constant current, 9.5 ampere, No. 7 Wood arc dynamo rated at forty 2000-candlepower lamps (Fig. 43). It was to supply 33 double carbon, Wood arc lamps connected in series and mounted on 40-foot-high cedar poles 300 feet apart along both sides of the canal. The outdoor arc lighting was designed to enable the canal to operate efficiently 24 hours a day (Fig. 49).[27]

In driving the generators, the speed of the turbines was regulated by vanes worked by electric governors.[28] When not under load running the generators, each turbine was regulated by a mechanical governor. This consisted simply of a bevel gear, mounted on the end of the horizontal turbine shaft, which meshed with a similar gear on the head of a vertical shaft extending down to the bottom of the flooded dewatering pump well in the basement of the powerhouse. Two propeller wheels were mounted on the vertical shaft which on turning in the water placed a constant load on the turbine shaft. The propellers were faced in opposite directions to cancel any thrust upwards against the bevel gears or downwards against the step bearing. When the mechanical regulator was not required, such as when the turbines were driving the generators or working the two dewatering pumps, the bevel gears were easily disengaged.[29]

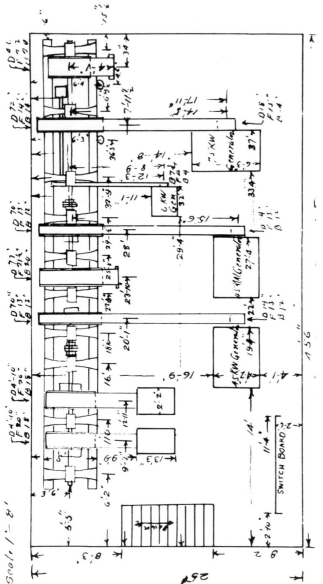

FIGURE 41. "Power House, Second Floor," showing drive pulleys on the main countershaft belted to (from l. to r.) the dual pump gearing system below the floor, the 45-kW power dynamo, the south turbine below, the 45-kW back-up power dynamo, a 75-kW arc lighting dynamo, and the north turbine below. The six-kW dynamo is belted beneath to the smaller countershaft of the auxiliary turbine system. (*Ross Notebook*, Sault Ste. Marie Canal Office, ca. 1930)

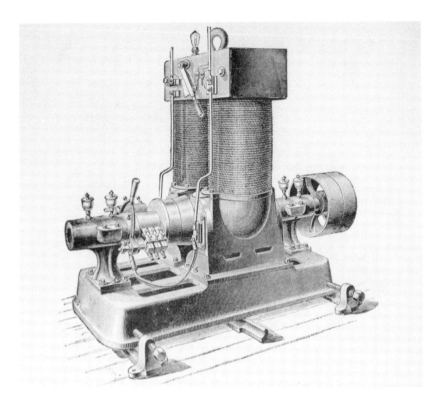

FIGURE 42. Edison bi-polar dynamo or "Long Waisted Mary Ann" with its two characteristically long, vertical field magnets. (Aitkinson, The Elements of Electric Lighting, 1890)

The dewatering pumps in the basement of the powerhouse were driven by the two main turbines using the same power transmission system as the generators. Two belts from the main countershaft on the second floor of the powerhouse drove a second horizontal shaft down on the first floor. A bevel gear on each end of this shaft meshed with a bevel gear on the head of a vertical shaft driving a dewatering pump in the pump well directly below (Figs. 38 and 41). The pumps were submerged in water entering the powerhouse through a culvert connecting with the discharge well of the floor sluice culverts in the lock chamber. The two 20-inch centrifugal pumps had a total pumping capacity of 32 000 gallons per minute, sufficient to dewater the lock in six to seven hours. They discharged into separate 20-inch-diameter pipes which merged into a single 30-inch-diameter pipe before emptying back into the canal below the lock. To drive the pumps, the two 45-inch turbines were coupled to provide 310 horsepower.[30]

FIGURE 43. Wood arc lighting dynamo, belt-driven through the pulley to the right. (Aitkinson, The *Elements of Electric Lighting*, 1890)

For indoor lighting, a third d.c. electrical generating system was installed independent of the two main turbines. Initially, it was intended to drive an incandescent lighting dynamo off the main countershaft.[31] Subsequently, however, a small 13-inch auxiliary turbine was installed to drive a three-kilowatt, 110-volt Edison bi-polar incandescent lighting dynamo. This turbine also was of the "New American" type on a horizontal shaft, with a 20-inch-diameter steel penstock connected to both branches of the main penstock above the 45-inch turbines. This by-pass was valved to enable the auxiliary turbine, and therefore the incandescent lighting system, to operate independently from either penstock branch, and to continue in operation even when both of the main turbines were shut down.[32] In the powerhouse there were 16 lights of 16-candlepower each on the incandescent lighting circuit, with additional lights to be added in a new machine shop next door.[33]

All three d.c. electrical systems in the powerhouse were controlled from a single switchboard seven feet high by five feet wide. It consisted of a heavy ornate frame of oak enclosing three panels of highly polished black slate two inches thick. In the centre panel were mounted the high voltage switches and meters for the two 500-volt electrical power generators with the 2000-volt arc lighting system switches and meters on the right panel and the 110-volt incandescent lighting system controls on the left panel (Fig. 44).[34]

The switchboard was of the latest design. During the 1880s, ammeters and voltmeters and various types of circuit breakers were invariably mounted on wood bases in any number of arrangements with all wiring connected to the front of the panels. But by 1890 a serious problem had emerged with the widespread introduction of street railways. The comparatively high voltage — 500 volt d.c. — had proved dangerous. All too often wood-based switchboard panels were set afire. As a result by 1892-93 a standard switchboard design had emerged. It consisted of panels of slate or marble, with all of the powerhouse meters and circuit breakers grouped on the front, and the dangerous live wiring rear-connected to enhance safety and improve the appearance of the switchboard. All of these advanced features, including the latest developments in meters and circuit breakers, were incorporated in the powerhouse switchboard on the Sault Ste. Marie Canal.[35]

In the powerhouse as of 1 July 1894, the scheduled completion date for the canal, the dewatering pumps were in place, and the installation of the two 45-inch turbines was well-advanced. The electrical power and arc lighting plant, however, was but half completed.[36] Moreover, work on the 13- inch auxiliary turbine for the incandescent lighting system did not commence until February 1895,[37] and a good deal of work remained as of that late date to render the lock gate and sluice valve machinery fully operational by electricity.[38]

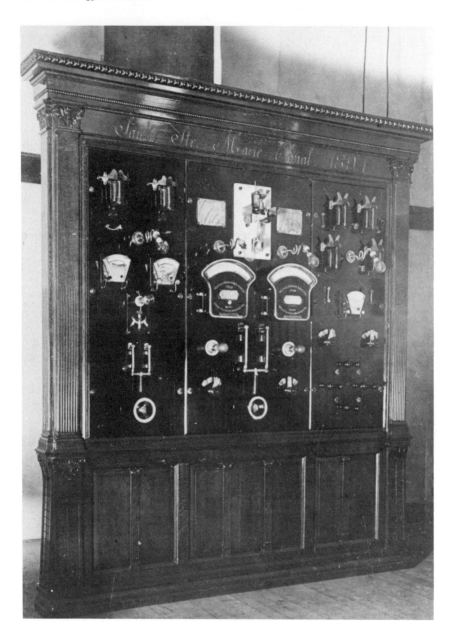

FIGURE 44. The powerhouse switchboard, with switches and meters for operating three d.c. electrical systems: the 110-volt incandescent lighting circuit on the left panel; the 500-volt power circuit in centre; and the 9.5-amperes arc lighting circuit on right. (Canadian Parks Service, Sault Ste. Marie Canal Office)

Lock Gate and Valve Operating Machinery

The specifications for the new electrically powered lock gate and sluice valve operating machinery had been ready for tendering as of 23 May 1894,[39] and the contract awarded to the Canadian Locomotive and Engine Works of Kingston, Ontario, shortly thereafter. But the correction of several design defects and problems encountered in manufacturing the novel machinery had caused innumerable delays.[40] The lock gate and sluice valve operating machinery were finally in place in October 1894, but only operational using their respective auxiliary manually operated cranks (Fig. 35). Indeed, the manual system of operation was used to lock through the steam tug *Rooth* on 15 October 1894 in the first test of the machinery.[41]

A total of six machines were required to work the lock gates — one for each leaf of the upper and lower gates, as well as for the auxiliary gates. Each machine was positioned on the lock wall coping adjacent to a gate leaf and contained two sliding pulley heads, one at either end of the machine to work a gate cable that passed down through an inclined well to the mitre post of a gate leaf. The heads moved horizontally back and forth on guide beams and worked in tandem. As one head was pulled inward on its travel, the other was pushed outward at the same rate of movement by the draw bars to which they were connected. This dual sliding pulley component of the new gate machine, as well as the arrangement of the gate cables, was identical in its design principle and layout to the abortive hydraulic system. The major change was in the means of moving the two sliding pulley heads.

In the new gate machine, hydraulic cylinders were replaced by an electrically driven, screw power system employing threaded screws to move the cross-head. The draw bars of the two sliding pulley heads were attached to a single cross-head with a horizontal travel. The cross-head was mounted on two four-inch-diameter threaded steel shafts in the screw machine. Both shafts were rotated by means of a spur gear system driven by a single electric motor. Driving the motor in one direction moved the crosshead horizontally along the threaded screws. This movement, by means of the attached drawbars, simultaneously drew in one sliding pully head and the cable it worked, while pushing outwards the other head to let out its cable. On reversing the motor, the cross-head moved an equal distance in the opposite direction, reversing the whole procedure. The cross-head had a travel of only eight feet nine inches, with a trip bar or switch stopping the motor at the end of each stroke (Figs. 45, 46 and 60).[42]

Only one stroke of the cross-head was needed to open, or close, a gate leaf owing to a four-part tackle arrangement of the cables in the gate machine frame. In each sliding pulley head there were two sheaves side by side, and the cable wound back and forth around the sheaves and a fixed snub pulley so that an eight

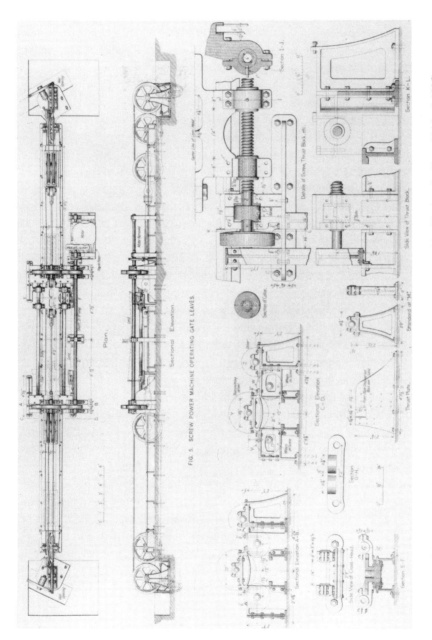

FIGURE 45. Drawing of horizontal screw power machine, for operating lock gate cables (*Engineering News*, 28 March 1895)

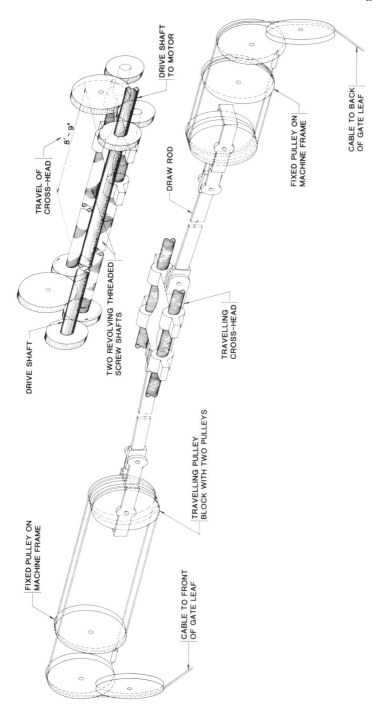

DRIVE SHAFT
TO MOTOR

TRAVEL OF
CROSS-HEAD

8' – 9"

DRAW ROD

FIXED PULLEY ON
MACHINE FRAME

CABLE TO BACK
OF GATE LEAF

DRIVE SHAFT

TWO REVOLVING THREADED
SCREW SHAFTS

TRAVELLING
CROSS-HEAD

FIXED PULLEY ON
MACHINE FRAME

TRAVELLING PULLEY
BLOCK WITH TWO PULLEYS

CABLE TO FRONT
OF GATE LEAF

FIGURE 46. Exploded view of a screw power machine, for working lock gates. (John Thomas, Engineering and Architecture Branch, Public Works Canada, 1987)

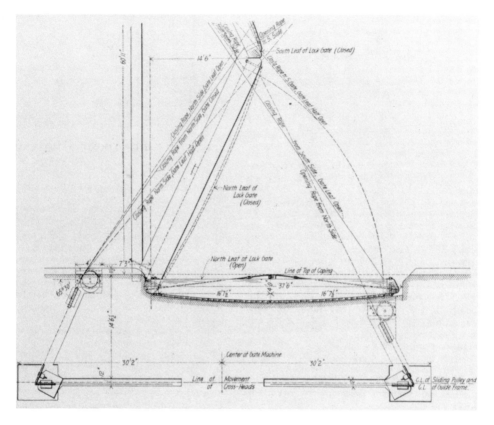

FIGURE 47. The original gate cables operating system. A single horizontal screw machine worked the opening cable of the adjacent gate leaf and the closing cable of the opposite gate leaf. (*Engineering News*, 28 March 1895)

foot nine inch travel by the sliding pulley head caused a 35-foot travel in the end of the cable attached to the gate (Figs. 46 and 47).

As initially designed and hooked up, each gate machine operated the opening cable of the adjacent gate leaf and the closing cable of the opposite gate leaf. The screw machine motors were wired to operate in pairs and synchronized in their operating cycle. In effect, to open a pair of gates, the opening cable of each leaf was drawn in by the adjacent machine while the closing cable of the opposite leaf was let out. For closing the gates, the procedure was reversed.[43] The synchronized system, however, apparently did not work as well as expected when tried. In June 1895, a pair of horizontal snub pulleys were positioned on the mitre sill at each pair

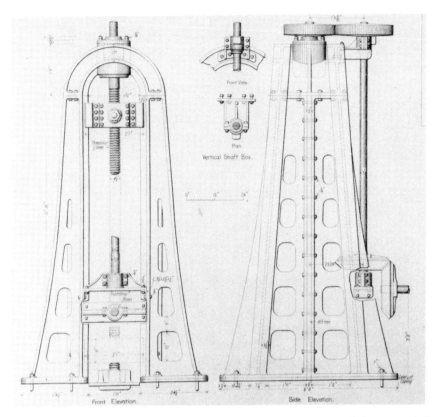

FIGURE 48. The vertical screw power machine, with travelling cross-head attached below to a vertical draw-rod and crank arm for pivoting the butterfly valves open and closed in the sluice culverts. (Engineering News, 28 March 1895)

of gates. Thereafter, both the opening and closing cables of each gate leaf were worked by the same gate machine (Fig. 52).[44]

To operate the butterfly valves on the floor sluice culverts of the lock chamber, a somewhat different type of screw power machine was designed. This machine had a single screw shaft and moved the cross-head through an up-and-down travel, rather than horizontally. A total of four valve operating machines were required, and each was positioned so as to straddle a three foot by five foot well passing down through the gate pier masonry of the lock (Fig. 48).[45] Once again an electrically powered screw machine was in effect simply substituted for the intended piston travel of an hydraulic cylinder. The crank and draw-rod system, designed previously for working the butterfly valves in the hydraulic system of operation, was retained unchanged.

In the lock chamber the eight steel butterfly valves — four on the filling culverts and four on the discharge culverts — pivoted on horizontal steel shafts of 10-inch diameter. Two valves were mounted on each shaft, to pivot in tandem, and the outer end of the shaft extended into a tunnel or culvert passing in through the base of the lock wall to a small vertical chamber. At the outer end of the valve shaft extension a crank arm of four-foot leverage was bolted, and attached, by a slotted connection, to a vertical draw rod. This 55-foot-long rod passed up through the three foot by five foot well in the gate pier masonry to the coping of the lock. There the rod was attached in turn to the cross-head of the valve screw machine (Fig. 32).

The cross-head had a vertical travel of 5.5 feet in the cast iron guides of the screw machine frame or standard. It was over 13 feet high overall with the screw shaft passing down through its centre. A single motor drove the 4.5-inch-diameter threaded steel shaft of the screw machine (Figs. 35 and 48). Driving the motor in one direction moved the cross-head upwards through its travel, and raised the draw-rod and crank arm, pivoting the two adjacent valve plates through a 90-degree rotation to a fully open position. When reversed, the motor drove the cross-head, and hence the draw-rod and crank arm, downwards pivoting the valves closed.

As with the gate operating screw machines, the motors of the valve machines directly across from one another on opposite sides of the lock were synchronized. This ensured that the four valves of the filling culverts would open and close in unison, and the four discharge culvert valves likewise. Similarly, the travel of each valve operating screw machine was terminated in either direction by an adjustable tripping bolt that on contact with the switch handle automatically cut off the current to the motor.[46]

A total of 10 motors — five paired units — were placed on the new Sault Ste. Marie Canal lock, but only four pairs of motors were required to operate the gate and sluice valve machinery during a lockage. The pair of motors operating the auxiliary gates were a standby unit to be used only when the lower gates required repair or were otherwise rendered temporarily inoperable. A heavily armoured submarine cable of 2.5-inch diameter connected each pair of motors across the bottom of the lock. All of the 25-horsepower motors were manufactured by the Canadian General Electric Company and were standard street railway car motors of their W.P. 50 series.[47]

The electrical components of the lock gate and sluice valve operating machinery were not fully operational until July 1895.[48] The following month, the electrical system of operation was tested as canal men were trained to work the machinery under the direction of J.C. Boyd, the new canal superintendent. As dredging continued on the approaches to the new canal,[49] several auxiliary buildings were constructed including six wood frame motorhouses to enclose the valve and gate screw machines on the lock wall coping, a workshop, an office building, and somewhat later a residence for the superintendent.[50] Finally, after a total ex-penditure of just over $3 296 000, the new Sault Ste. Marie Canal was completed, the last link in an all-Canadian inland navigation stretching over 1400 miles from the head of Lake Superior to the ocean port of Montreal. The official opening was held on 7 September 1895 when the new Canadian passenger steamer, the *Majestic*, passed through the lock with 700 passengers and several dignitaries on board.[51]

"a quickness of despatch"

The electrically powered lock operated smoothly from the beginning and proved far superior to the hydraulically powered Weitzel lock of 1881, and to the new Poe lock opened subsequently in August 1896, on the American St. Mary's Falls Ship Canal. It also confirmed the wisdom of opting for the long lock, passing vessels one astern of another, rather than the more traditional flotilla lock configuration of the American locks with their comparatively broad chambers for passing vessels two abreast (Fig. 49).

FIGURE 49. "Three in Lock, First Day," 7 September 1895. An enclosed arc lamp is in foreground, as well as a motorhouse built to enclose a vertical and horizontal screw power machine. (Canadian Parks Service, Sault Ste. Marie Canal Office)

In the time required for filling the lock chamber during a lockage, or empty-ing it, there was little difference between the Canadian lock and the new Poe lock. Both had basically the same floor sluice culvert system which enabled water to be let in, or out, of the lock at a much faster rate than on ship locks with the conven-tional side-wall culverts or the gate sluice system. But there were still limits beyond which excessive turbulence would cause surging presenting the danger of a ship striking and carrying away a lock gate. Consequently, it took about seven minutes to fill the Canadian lock in locking up, and five minutes to empty it in locking down — a rate just slightly better than the new American lock.

The superiority of the Canadian lock rested in the instantaneous operation of its electrical power system and the despatch with which large vessels could enter and depart from the long lock. On the Canadian lock, it took only 50 seconds to close the gates, and 50 seconds to open the butterfly valves; whereas two minutes were required just to open, or close, the gates on the hydraulically operated locks. Moreover, in the American flotilla locks, tugs were required to place the vessels. This entailed a major time loss, with an attendant danger of vessels surging into the lock walls as they were manoeuvred into place. In the Canadian lock steam ves-sels were able to proceed through under their own power. Even steamers towing sailing schooners proved able to take their flotilla through without the aid of a tug. In sum, it took anywhere from 36 to 40 minutes for vessels to lock through either the Poe or the older Weitzel lock; but vessels could lock up through the Canadian canal in as little as 14 minutes and 14 seconds, and lock down in two minutes less (Fig. 50).[52]

The New Electrical Power System

Although the employment of electricity to operate a canal lock was unprece-dented, no problem was experienced with the electrical power system.[53] The ap-plication was unique, but it made use of an established technology of proven reliability where generators and motors were concerned.

Dynamo-electric generators had been developed in Europe during the 1860s,[54] and introduced into commercial production in North America a decade later. In 1878, the Brush Electric Company of Cleveland, Ohio, began marketing an arc lighting system with a reliable dynamo, a current regulator, and a newly improved carbon arc lamp capable of running for hours on end. This had marked the begin-ning of commercial electric lighting in the United States, and within five years a number of arc lighting companies were in business, each with a dynamo and arc lamp of its own design.[55] Unlike earlier generators, the dynamo-electric generator or dynamo could be built to produce almost any voltage or amperage desired.[56]

FIGURE 50. Lake boat leaving Sault Ste. Marie Ship Canal lock. (K. Elder, postcard collection)

But the first of the arc lighting dynamos were on a small scale, operating only four to 16 arc lamps in series, with an efficiency of no better than 38 percent.[57]

The real potential of the dynamo had only become apparent in 1879-80 when the newly formed Edison Electric Light Company introduced the Edison bi-polar generator as part of Thomas Edison's new incandescent lighting system. These dynamos, called "Long Waisted Mary Ann's" because of the characteristic long vertical cylinders of its two field magnets, had a claimed efficiency of a phenomenal 90 percent, and set the standard for dynamos developed thereafter. The direct current bi-polar dynamo was manufactured in four sizes and rated according to the number of 16-candlepower incandescent lamps on a circuit: namely a 15-lamp generator, a 60-lamp, a 150-lamp, and a 250-lamp generator (Fig. 42).[58]

For exhibition purposes, the Edison company had also built in 1881 the first large scale — in contemporary terms — generator. The 200-horsepower "Jumbo" generator, which weighed 27 tons, proved capable of running twelve hundred 16-candlepower incandescent lamps connected in parallel. It had been displayed at a London Exhibition, and then installed in a plant at Holborn Viaduct where it was used, on commencing operation in March 1882, to light surrounding public buildings.[59]

Subsequently, six 100-kW Jumbo generators, each of about the same capacity as the original and capable of producing a direct current of 850 amperes at 115-120 volts, were installed in a New York power plant. When started up in September 1882, with only one generator then in operation running 400 incandescent lamps, the Pearl Street Station was the first central power station for incandescent lighting in North America. Within a year its generators were operating a total of over 8000 lamps in hundreds of offices within roughly a square mile area.[60] Other electrical manufacturing companies followed suit in building comparatively large direct-current generators, and within a decade 200-kW dynamos were being manufactured and in common use — principally for powering electric street railway lines.[61]

When the canal engineers at Sault Ste. Marie opted for an electrically powered lock in 1893, the direct-current system of electric lighting, both incandescent lighting for indoors and arc lighting for street lighting, was well-established in North America. Electrical power, in contrast, was in its infancy. It had yet to come into common use in the factories and workshops of North America, but was widely employed in powering electric street railways.

Since 1873 North American engineers had known of the principle of dynamo reversibility: that when two dynamos were connected the current from the first would cause the armature of the second to rotate with sufficient power to drive a shaft or axle. Thereafter efforts to develop efficient direct-current motors had paralleled generator development and culminated in their being introduced into commercial use for powering streetcars.[62]

In the period 1880-85, several American electric lighting companies, and newly formed electric manufacturing companies, had begun experimenting with electrifying street railways. Trials of electric streetcar prototypes were made at Edison's Menlo Park (1881-82), in Chicago (the winter of 1882-83), at Saratoga (1883), at the Canadian National Exhibition in Toronto (1884), and Cleveland (1884-85), among others. None of these temporary experimental lines, however, were more than two-miles long, and all had employed the existing technology of the electric lighting industry. The motors consisted simply of smaller lighting dynamos, ranging from eight to 12 horsepower and operating on roughly 110-112 volt circuits.[63] These experimental lines had been plagued by motor problems and breakdowns, which had discouraged the spread of electrification. As late as the spring of 1887, there were as few as eight electric street railways in the United States with none having more than eight to 10 electric cars, and many engineers had still believed electric traction to be totally impractical.[64]

The basic problems involved in applying electrical power to running street railways, however, had been solved within three years, 1885-88. Several companies developed large, double reduction motors of 15 horsepower for streetcar applications. These motors had an efficiency of up to 90 percent and proved highly durable once the critical problem of excessive wear in the copper brushes were eliminated through substituting carbon brushes. Controls had also been developed for operating more than one motor simultaneously from a single source, and to suppress arcing caused by the heavy surge of current required on commencing steep grades. Moreover an adjustable split-gear drive system was developed which, with the introduction of spring-suspended, dual motors mounted on trucks beneath the cars, ensured a proper meshing of the drive gears at all times in moving forward or backward, thereby preventing jerky motion and/or the stripping of gear teeth.

These and other technical developments, in conjunction with the manufacture of large 500- to 600-volt electrical power dynamos, had made possible the construction of dependable electric street railway lines of up to a dozen miles in length. As of the autumn of 1888, engineers were convinced that the new technology for applying electrical power to street car propulsion had proven itself, and were predicting that electric street cars would rapidly render horse-drawn cars obsolete.[65] A virtual boom in electric streetcars had followed as cities in both the United States and Canada rushed to introduce electric traction in place of horse-drawn vehicles.[66]

Prior to 1888, only two street railways were operating in Canada with electrically powered cars: a two-car, two-mile long line in Windsor, Ontario; and an eight-car, eight-mile line in St. Catharines, Ontario.[67] But as of 1895, electric streetcars had been introduced into as many as a dozen major Canadian cities, with several street railways having converted completely to electric traction.[68] The streetcars had proven, at first hand for Canadian engineers, the reliability of direct-current motors in standing up to the constant starting and stopping that would be required in operating lock gates and valves, and had shown that electric motors were unaffected by freezing temperatures.[69]

In view of the existing state of electric street railway technology, and Thomas Munro's practical experiments in operating a lock gate with electrical power, the 1893 decision to convert the Sault Ste. Marie Canal to electrical power did not constitute a major technological advance. Indeed, although the application was novel, the ten 25-horsepower motors, and the split gear adopted for connecting the motor shafts to the drive gears of the screw machines for operating the lock gates and floor sluice valves, were standard items of contemporary electric street railway technology.[70] Nonetheless, the electrification of the Sault Ste. Marie Canal lock was a significant undertaking where canals were concerned and attracted a good deal of attention with inquiries being received from as far away as the Netherlands.[71] It was, however, but one of many new applications in an era of transition where electrical power was concerned.

Efficient and compact d.c. motors had been developed during the 1880s for operation in workshops and factories on low voltage incandescent lighting circuits. Initially, motors ranged between one-sixth to one horsepower, but by 1885 a 15-horsepower motor was introduced for industrial applications, and in 1890 a 30-horsepower motor. By that date, direct-current motors were being used in a variety of applications running sewing machines, freight elevators, cranes, lathes, drills, planers, and saws as well as pumps and rock drills.[72] Indeed, an electrically powered rock drill had been experimented with in excavating the Sault Ste. Marie Canal.[73]

As early as 1888, North American engineers had begun to regard the electric motor as being far superior to small steam plants for driving machinery. Electric motors were quieter, cleaner, and more efficient. They could be started instantaneously, and turned off when a machine was not in use. In contrast, steam plants had to be fired up before machinery could be activated, and demanded constant attendance with a good deal of labour involved. Where factories were already using electricity purchased from a central station, small electric motors were cheap to install and operate on the lighting circuit. Where fire hazards were not extreme from motor sparking, the motors could also be placed adjacent to the machinery to drive an individual machine or a group of machines. This saved on the cost of installation and power loss involved in driving a complex system of line shafting, pulleys and belting from a steam plant a distance removed from the machinery.[74]

Despite the obvious advantages of electrical power for a wide variety of applications, electric motors had not been widely installed in North American factories and workshops prior to 1893. Where isolated private power plants were concerned, there was little incentive to change. The power plant still had to be maintained whether the steam engines were used to drive line shafting or dynamos. Even in factories where hydraulic power was being used, converting machinery was still a costly undertaking. Moreover, the direct current system of transmission was not without its drawbacks where electrical power was concerned.

This had become apparent after 1890 with the first major conversion of a textile mill — the Ponemah Mills at Taftsville, Connecticut — to electrical power. Voltage drop had proved a serious problem affecting motor speeds on the low voltage incandescent lighting circuits. Turning on the lights at dusk lowered the voltage, decreasing motor speeds. But increasing the voltage to compensate for the lights, caused the motors to speed up when the lights were turned off. In effect, direct-current motors on lighting circuits had proved ill-suited to applications where a constant speed was required, as in running textile machinery.[75] In the new Sault Ste. Marie Canal undertaking, however, the same factors did not apply.

At Sault Ste. Marie, hydraulic turbines rather than steam were being used to drive the dynamos, and the engineers had obviously taken into account the shortcomings of the direct-current system in designing the electrical power system for operating the lock gates and sluice valves. The transmission circuit was short, owing to the powerhouse being adjacent to the lock chamber. But voltage drop was further minimized by running only the motors off the main generators, and operating the power circuit at the comparatively high voltage of electric traction power systems — 500 volts — with separate generators at different voltages for the arc and incandescent lighting systems.

These precautions obviously were responsible for the lack of any transmission problems where the electric motors were concerned. The change in the lock gate cables system, to run both cables of each gate leaf from the same screw machine, however, may well be indicative of a problem in synchronizing the speed of opposite pairs of motors. If the voltage dropped appreciably in passing through the submarine cable to the motor on the opposite side of the lock, its speed would have been slowed relative to the motor on the near side adjacent to the powerhouse. This would have placed a severe strain on the gate cables in the original system where each motor had operated the opening cable of one gate leaf and the closing cable of the opposite gate leaf.

Obtaining 25-horsepower d.c. motors for the 'Soo' Canal installation posed no difficulty. Indeed, on receiving the contract for the canal electrical systems in May 1894, the Canadian General Electric Company — formerly the Canadian

Edison Manufacturing Company — was well situated to supply the motors, as well as the machinery for the electrical power and lighting plants.

The Edison Electric Light Company of Canada had been established in Hamilton, Ontario, as early as May 1882 to manufacture incandescent lamps for the Edison lighting system. Dynamos were purchased from a licensed manufacturer in Sherbrooke, Quebec, until 1888 when the plant was taken over by the newly incorporated Canadian Edison Manufacturing Company, and moved two years later to Peterborough, Ontario. There the manufacturing line was expanded to include motors for industrial and electric traction applications, electric wiring, and underground conductors as well as dynamos. This capacity had been increased still further in 1892 when another Canadian company supplying Thomson-Houston electrical equipment — the Toronto Construction and Electrical Supply Company — was absorbed on the forming of the Canadian General Electric Company (C.G.E.).[76] By that time, C.G.E. had gained access, through a series of major mergers involving its parent company in the United States, to almost a decade of technical experience in the development of d.c. motors and to an even older, and more varied, technology involving various arc lighting systems.

Neither the Canadian Edison Manufacturing Company nor its American parent had been involved in the early development of the direct-current motor. Other companies, principally the Sprague Electric Railway and Motor Company and the Thomson-Houston Electric Company, had taken the lead in introducing and improving the d.c. motor. The Edison Electric Light Company and its Canadian subsidiary had concentrated on marketing the Edison incandescent lighting system,[77] but in the United States an Edison company had been involved from an early date in manufacturing d.c. motors.

From the founding of the Sprague company in November 1884, Sprague motors had been manufactured under contract by the Edison Machine Works of New York, and in 1889 the Sprague company had been absorbed on various Edison companies being consolidated to form the Edison General Electric Company. This take-over had been followed in 1892 by a mammoth merger with the Thomson-Houston Company to form a new entity — the General Electric Company — producing a wide range of electrical equipment for d.c. power and lighting applications, single phase alternating current (a.c.) lighting and henceforth also polyphase a.c. applications.[78]

The Thomson-Houston Company was one of the largest electrical manufacturing companies in the United States with production facilities on a scale comparable to Edison General Electric. Prior to the merger, Thomson-Houston was among the leading manufacturers of d.c. motors and electric street railways, and by far the leading manufacturer of arc lighting systems. Founded in April 1883 to produce the Thomson-Houston arc lighting system, the company had added the production of incandescent lighting equipment in 1884, d.c. railway motors in 1886, single phase a.c. lighting components in 1887, and as of 1889 had begun manufacturing a complete system for d.c. electric street railway installations.

An active policy of acquisitions had brought the absorption of a major electric street railway manufacturer — the pioneer Van Depoele Electric Company in 1888 — and by 1890 resulted in Thomson-Houston controlling the patents and production facilities for three of the leading arc lighting systems in the United States: the Thomson-Houston; the Brush; and the Wood.[79]

As of 1894, two years after the formation of the General Electric Company, its Canadian subsidiary — Canadian General Electric — was already manufacturing 25-horsepower electric street railway motors, as well as the 45-kW, 500-volt Edison bi-polar dynamos required for the Sault Ste. Marie Canal power installation. C.G.E. was also well-established as a manufacturer of the Edison incandescent lighting system: dynamos, lamps, wiring, and switchboards. All of this equipment was manufactured in Peterborough. Only the arc lighting component of the Sault Ste. Marie Canal contract required C.G.E. to go outside the traditional Edison line of its own manufacture. Here the Wood system was selected: the No. 7 Wood arc dynamo (Fig. 43), and the 33 double carbon, enclosed Wood arc lamps.

Where the electrically powered lock was concerned, the real innovation was in designing the screw machines and switches for operating the gates and valves by electrical power. Here the only problem experienced was structural rather than electrical. The long draw rods of the valve crank arm proved of too light a construction with a tendency to bend. After one or more had been straightened several times, all of the draw-rods were taken out and strengthened during the winter of 1897, and then replaced three years later with heavier rods.[80] Ironically, real problems were experienced with the two electric lighting systems which employed well-established technologies in a conventional manner.

The New Lighting System

At the time the Sault Ste. Marie Canal electrical plant was designed, contemporary engineering practice called for two distinct lighting systems — arc lighting for outdoors; and incandescent lighting for indoors. But when first introduced into Canada both systems had been used indiscriminately for either purpose.

As early as 1879, arc lighting had been introduced in a Toronto restaurant,[81] and in June 1881 the Brush company had installed a 40-arc lamp circuit — probably the first major electric lighting installation in Canada — for illuminating the factories, workshops, mills, streets and yards of the E.B. Eddy saw mill complex at the Chaudiére Falls in Ottawa.[82] Subsequently, arc lamp systems had been introduced for street lighting, commencing in December 1882 in Montreal and spreading within a decade to over a dozen Canadian cities.[83]

Incandescent lighting had also been introduced into Canada at a very early date, primarily for lighting the interior of textile mills where good lighting was of a critical importance.[84] In the first major installation, completed in February 1883, five hundred 16-candlepower lights on the Edison system were employed to illuminate a new 500 foot by 120 foot weave shed of the Canada Cotton Manufacturing Company at Cornwall, Ontario. This had been followed in September 1883 by an 800-light Edison system in the Montreal Cotton Company mills at Valleyfield, Quebec, and in January 1884 by a somewhat smaller circuit — soon increased to 1000 lights — in the Parliament Buildings at Ottawa. Few difficulties were encountered in introducing electric lighting to factories and public buildings, in contrast to the problems that had plagued the introduction of electric motors for powering streetcars. As a result, by 1890 there were over nine private installations of 600 or more lights on a circuit. In a number of towns private companies operated central stations supplying stores, offices, and public buildings with incandescent lighting. Domestic lighting had scarcely been introduced.[85]

Incandescent lamps had been installed in several Canadian cities for street lighting and for lighting a canal. In January 1887, the first incandescent central station had opened in Victoria, British Columbia, followed by Vancouver in September 1887 and Valleyfield, Quebec, in January 1888. The same year, some 160 Edison incandescent lights, each of 50 candlepower, were mounted on poles 150 feet apart to illuminate the entrance basin and first three locks of the Lachine Canal in Montreal. Incandescent lighting, however, had proved less than satisfactory for outdoor use. On the Lachine Canal, and in Vancouver, arc lighting systems were soon substituted.

These experiences had convinced Canadian canal engineers that the soft glow of incandescent lamps was too dim for outdoor or street lighting, but ideal for indoors. To the contrary arc lamps, which cast a brilliant light up to 400 feet in extent, were adjudged better suited for street lighting, but had too bright and harsh a glare for indoor use.[86] In the United States, this had been recognized much earlier. As of the mid-1880s major arc lighting companies had begun to add incandescent lights to their systems, and incandescent lighting companies began using arc lamps.[87] Initially, incandescent lamps were grouped in parallel on the series circuit used for arc lamps. But the best arrangement proved to be two separate systems with arc lamps in series on a high voltage circuit — either 2000 or 3000 volts — for street lighting; and incandescent lights on a parallel circuit of 110 volts for indoor lighting.[88]

Where electric lighting on Canadian canals was concerned, the separate functions of arc and incandescent lighting were well established prior to electric lighting being installed on the new Sault Ste. Marie Canal. On the Lachine Canal, the arc lighting system which replaced the unsatisfactory incandescent system, had arc lamps along a section of the canal with incandescent lamps, on the same circuit, for lighting nearby workshops and storage sheds.[89] In 1891, electric lighting had also been introduced to the Chambly Canal at Chambly, Quebec on the Richelieu River. There two separate generators and wiring systems were installed in an hydro-electric powerh:n site. One circuit ran arc lamps along one side of the canal for about 1.5 miles, and the other circuit operated incandescent lights in the workshops.[90] The 'Soo' canal arc and incandescent lighting installations were by no means unprecedented, but nonetheless proved far from trouble free.[91]

Initially a problem was encountered with the small 13- inch auxiliary turbine installed to drive the incandescent lighting dynamo. The turbine proved too power-ful for its work.[92] In North America, in contrast to the European practice, turbines were not designed for a particular application. They were purchased off-the-shelf from turbine manufacturers in whatever size appeared appropriate. The speed of the turbine shaft was then modified by calculating the pulley arrangement required in the belt drive to produce the speed suited to a particular generator. Where tur-bine performance was concerned, American companies had always followed a "cut and try method" whereby the buckets or vanes of a turbine were modified after in-stallation under actual working conditions to improve efficiency or attain a desired output. Even the characteristic design of the highly efficient "American" turbine had evolved in this manner.[93]

This traditional North American approach to turbine performance was fol-lowed at the 'Soo'. The manufacturer of the small turbine, William Kennedy & Sons, simply suggested that three of the six gates of the turbine be closed, and the altered turbine tried out. If this solved the problem, then the gates were to be per-manently fastened shut.[94] This was apparently done, but then another problem emerged. The three-kW incandescent lighting dynamo proved too small for run-ning a total of 22 lamps of 32 candlepower in the powerhouse and workshop.

To remedy this problem, the incandescent dynamo was disconnected and the incandescent lighting circuit connected to the 9.5-ampere arc lighting dynamo sys-ten. This dynamo was rated at 40 lamps of 2000 candlepower. It already ran 33 arc lamps to illuminate the canal at night and was in use day in and day out for as many as 14 hours at a stretch during the dark rainy days and nights of autumn. The 22 in-candescent lamps added to the circuit were of a comparatively low candlepower, but the heavier load overtaxed the arc dynamo. Over the first two years of opera-tion, seven coils were burnt out on one armature of the arc lighting dynamo, and two coils on another armature. The length of the lighting circuit was a matter of serious concern. On the expanded lighting circuit there was over seven miles of wire including 4250 feet of submarine cable.[95]

Where d.c. electricity was concerned, energy loss was a critical problem whether in point to point transmissions or in distribution circuits. At an early date experimental long-distance, point-to-point, transmissions had been made in Germany and France at comparatively high voltages. In 1882, power was transmitted at 2700 volts over 34 miles of iron wire to Munich, and in 1886 at 6000 volts over a 36-mile line to Paris. But over half the energy was lost in transmission, achieving an efficiency of only 32 and 45 percent respectively.[96] Despite improvements in wiring, including the substitution of copper for iron wire, long distance transmission of d.c. electricity had remained totally uneconomical, and virtually non-existent.[97]

To avoid excessive power loss, direct-current power plants — whether isolated factory installations or central stations — were situated right at the distribution system.[98] To that end, steam engines were generally employed to drive the generators. Hydraulic turbines were used only where water power was available within a few miles of the distribution centre or directly on site. This was the case with the Sault Ste. Marie Canal and the several canals where electric lighting had been installed previously.[99] The power loss problem, however, could not be avoided in distribution circuits and the limiting factors were much greater.

Power loss in a direct-current system could be reduced in two ways: by increasing the voltage so as to proportionally reduce the amperage or current flow; or by increasing the carrying capacity or gauge of wire to facilitate current flow. But high transmission voltages heavily increased insulating costs and presented a serious element of danger. With arc lighting dynamos of 40 and 60 lamp capacity, producing 2000 and 3000 volts, armatures all too often blew out as if struck by lightning.[100] Moreover, direct-current electricity had to be used at generator voltage which ruled out high voltage distribution circuits for incandescent lighting, and many arc lighting applications. Where the carrying capacity of wiring was concerned, the cost of copper was the critical factor. There were definite limits to the length of circuits beyond which the cost of the amount of copper required to eliminate excessive power loss, and variations in voltage, was prohibitively expensive.[101]

As early as 1883, Edison had introduced a three-wire feeder system of distribution that reduced the amount of copper wiring in an incandescent lighting circuit by as much as 60 percent.[102] But these 110-volt lighting circuits were still uneconomical beyond a half-mile radius of distribution from the central station.[103] For the comparatively higher voltage — 500 to 550 volts — of electric traction systems, distribution circuits were only economically feasible for up to five or six miles. Street railway powerhouses were commonly placed so as to supply sections of only three or four miles in extent in any direction.[104] The effective circuit of the standard 2000- and 3000-volt arc lighting systems ranged beyond these limits, depending on the number of lamps and the total resistance in the circuit.

With canal lighting, the major difficulty was in working with a large number of arc lamps spread out along an extensive series circuit at comparatively great distances apart. Conveying and regulating d.c. electricity over such a circuit was much more difficult and limiting than in a point-to-point transmission.[105] From a relatively early date, arc lighting had been considered for illuminating major canals for nighttime operation, but the great distances involved had precluded or severely limited such applications. On the several canals where electric lighting had been introduced, only short sections were illuminated such as at a major lock, set of locks, or at a basin/harbour entrance to the canal.[106]

The Sault Ste. Marie Canal was the first canal in North America to be totally illuminated with electric lighting.[107] This was possible, with direct-current arc lighting, only because of the canal's comparatively short length — just over a mile, exclusive of the approaches that were also illuminated. Nonetheless, it was quite an extensive lighting system for the voltage employed, with its 33 arc lamps spaced on a series circuit at 300-foot intervals along both sides of the canal. The expedient of adding the parallel circuit of 22 incandescent lamps to the arc lighting system, could only have exacerbated the lighting problem.[108]

d.c. or a.c.?

In response to the deficiencies in the lighting system on the Sault Ste. Marie Canal, consideration was initially given to replacing the direct-current electrical plant — both the lighting and power generation systems — with alternating-current machinery. Inquiries were made of the Canadian General Electric Company concerning the size of turbines required to drive a.c. dynamos or so-called alternators as large as 200 kW and 250 kW, and the speeds and speed reduction ratios of a.c. induction motors of 20, 25 and 30 horsepower and the new C.G.E. 800 motor.[109] Various arrangements were obviously being considered for operating a polyphase a.c. system with new motors — either a.c. or d.c. — and both lighting circuits being run off a single electrical system utilizing alternators, transformers, and perhaps a rotary convertor. The 800 motor had been designed for d.c. street railway applications, but could operate on a polyphase a.c. system if a rotary convertor were used. As of November 1897, however, plans changed.

The canal engineers began to plan for converting only the lighting system to alternating current. The new plan called for the installation of a large horizontal shaft turbine, supplied with an eight-foot-diameter penstock, and two alternating current dynamos to be direct-connected, with a clutch coupling, to either end of the turbine shaft. These dynamos, one presumably a back-up, were to run only the lighting system — transmitting at a high voltage with transformers to step down the voltage for use by the lamps.[110]

Owing to the heavy cost involved in converting the lighting plant to alternating current, the canal engineers ultimately decided to simply maintain the direct-current system. The dynamo capacity was increased to the maximum extent possible with the existing arc lamps. A No. 8 Wood arc lighting dynamo of 9.6 amperes, with an automatic current regulator and 60-lamp capacity, was purchased from the Canadian General Electric Company. In effect, the arc lighting system voltage was increased from 2000 to 3000 volts.

As of February 1898, the new generator was placed in the powerhouse next to the existing 40-lamp, No. 7 Wood arc lighting dynamo which was retained as a back-up. Both dynamos were driven by belting off the countershaft of one of the existing 45-inch turbines.[111] At the same time, the incandescent lighting system was re-established on its separate 110-volt circuit.

Initially, consideration had been given to purchasing a three-kW, shunt-wound, Edison bi-polar, incandescent lighting dynamo to be driven by the 13-inch auxiliary turbine in the powerhouse. Then the feasibility of installing a six-kW dynamo was investigated. The larger dynamo, however, required 10 horsepower to operate efficiently, and the auxiliary turbine — taking into account the 18-foot head of water, friction loss, and an elbow in the discharge tube that constricted the flow of water exiting from the turbine — produced an estimated output of but slightly more than 10 horsepower at the dynamo. Rather than taking a chance on "running pretty close," it was decided simply to restore the original three-kW dynamo to operation, despite its earlier deficiencies.[112] In effect, the indoor incandescent lighting system would run henceforth with all the lamps producing a relatively poor light, or some of the lamps would have to be disconnected to increase the current to, and thus the brightness of, the remaining lamps on the circuit.

Conversion costs had ruled out any changeover to an alternating-current system to solve the lighting problems on the Sault Ste. Marie Canal. But by all indications, as of 1897-98, the engineers clearly preferred the a.c. system for extended canal lighting installations. They were also willing to consider either operating d.c. motors off of an a.c. system, employing a rotary convertor, or replacing these motors with a.c. motors. This preference for running both lighting and power applications off an alternating-current system was a reflection of a series of on-going technical developments that had culminated in the years 1893-95 when the Sault Ste. Marie Canal was being electrified.

Alternating-current electricity had co-existed with direct-current systems since the development of the dynamo in the 1860s. All dynamos produced a form of alternating current, but the earliest arc lamps had required a continuous current to operate the regulator controlling the feeding mechanism of the two carbon rods. Hence commutators were placed on dynamos for converting the alternating current to direct current for arc lighting applications, and this system had come to prevail in North America from the first introduction of arc and incandescent lighting. Several types of early arc lamps were manufactured for operation on alternating current, but a.c. lighting systems had remained insignificant in North America until the introduction of the transformer.[113]

In 1882-1883, several experiments in long-distance transmission had been conducted in England by Lucien Gaulard and John D. Gibbs employing induction coils, or what became known as transformers, on an alternating-current circuit. In one experiment, an a.c. current at 500 volts was transmitted 16 miles over a series circuit with on-line transformers that reduced the voltage to 50 volts for the operation of incandescent lamps on secondary circuits. Transformers had soon proved up to 90 percent efficient, and capable of producing any desired voltage in secondary consuming circuits.[114] In effect, all of the advantages of high voltage long-distance transmission could be obtained, and yet the electricity readily distributed at lower voltages. This was a decided advantage over direct-current electricity which could not be transformed. It had to be used at generator voltage. The commercial potential of the transformer as a voltage-changing device, however, had first been grasped by the Westinghouse Electric Company of Pittsburg, Pennsylvania.[115]

The Westinghouse company had improved the transformer and alternating-current generator (alternator), and had an incandescent lighting system on the market by 1887 with a standard transmission voltage of 1000 volts in a primary parallel circuit stepped down by transformers to 50 volts for lamps on the secondary parallel circuits. The Westinghouse a.c. system proved not only superior to direct-current for long-distance transmission, but the lower consumption voltage actually resulted in incandescent lamps having a greater efficiency and longevity. Within two years, almost all of the major electrical manufacturing companies had come out with an a.c./transformer system of incandescent lighting in addition to their existing d.c. lighting systems. By 1889 most of these companies, including the Westinghouse company, were also manufacturing arc lamps for use on the various a.c./transformer systems.[116]

During the period 1887-1892, electric lighting had spread widely throughout North America with a fierce rivalry developing between the Westinghouse company, the chief proponent of the a.c. system, and Edison General Electric, the one major company that continued to adhere exclusively to the d.c. system. Each system, however, had at least one significant drawback: there was no practical motor capable of operating off the single phase a.c. system used for lighting; and the transmission/distribution limits of the d.c. system made it economical only in densely populated areas or in isolated factory installations with the powerhouse relatively close by. Consequently, the new a.c. lighting systems had spread for the most part to towns and villages where the population was relatively dispersed. In the United States, it was also employed in several long distance-transmission systems — up to distances of 35 miles at 10 000 volts — where a major water power site was comparatively far removed from the distribution area.[117]

In Canada, no long-distance a.c. transmission systems were built prior to 1894, but single phase a.c. lighting systems were well-established. In 1888, the first a.c. plant had been built in Cornwall, Ontario,[118] and in the same year, another established at Thorold, Ontario — on the Welland Canal where a 12-foot head drove a turbine belted to a 120-kW, single phase a.c. generator producing 2400 volts.[119] Two years later, there were over 147 central stations in Canada generating electricity. Most were d.c. stations for arc street lighting, and to a lesser extent for d.c. incandescent lighting. But the number of a.c. incandescent lighting stations was growing rapidly. Out of a total of about 70 000 incandescent lamps in use throughout the country, 23 500 were on the Edison d.c. system, 6850 on the new Westinghouse a.c. system, and the remainder spread among manufacturers who produced both d.c. and a.c. incandescent lighting systems.[120] But as of the summer 1893, when the decision had been made to employ electrical power for operating the new Sault Ste. Marie Canal, all electrical power was supplied on the d.c. system. The technology existed for building an a.c. power system, but the equipment required was not then commercially available.

As early as 1888, Westinghouse had begun work on polyphase a.c. motors developed and patented by Nikola Tesla. These motors, however, did not run well on the high frequencies — 100 cycles per second and more — used in contemporary single phase a.c. lighting systems. They also had a low starting torque and consequently were totally unsuited for most major power applications, such as propelling electric street railway cars. But in the early 1890s a breakthrough had been achieved. When operated on two- or three-phase a.c. systems of low frequency — 30 cycles per second — a.c. motors proved as proficient as d.c. motors with several additional advantages. They were less susceptible to speed variations, caused in large part by voltage fluctuations, and the absence of a commutator not only eliminated all sparking problems but enabled a.c. motors ultimately to be constructed on a far larger scale than d.c. motors.[121]

Within a one year period, 1893-94, three major companies — Westinghouse, General Electric, and the Stanley Electric Manufacturing Company — had begun constructing low-frequency polyphase a.c. power systems, and marketing a.c. induction motors as large as 110 horsepower for factory installations. Among the earliest users had been textile mills. The non-sparking feature of the induction motor enabled them to be placed directly in the mill, eliminating line shafting and extensive belting, and the constant speed characteristic enabled both light and power to be supplied off the same generator.[122]

The real flexibility, and ultimately superiority, of the polyphase a.c. system was first made apparent at the Westinghouse display at the Chicago World's Fair in 1893. There transformers, induction motors, and the newly invented rotary converter for converting a.c. into d.c., were incorporated into a single lighting and power system. Large generators in a central powerhouse produced two-phase a.c. voltage which transformers stepped up for transmission, and then stepped down to various voltages for distribution to arc and incandescent lighting circuits and power circuits running a.c. motors. At several distribution points, rotary converters changed the a.c. voltage into d.c. to operate railway motors and other d.c. applications. In effect, a universal system of electric power and lighting had been inaugurated which had its major impact subsequently at Niagara Falls, New York.[123]

When the American Niagara Falls hydro-electric plant opened in August 1895, it far surpassed any previous electrical power and/or lighting plant. In the initial installation, three vertical shaft turbines of 5000 horsepower were employed with each driving a 5000-horsepower, two-phase a.c. Westinghouse generator producing 2200 volts. At the powerhouse, 1250-horsepower transformers converted the current to three-phase and stepped up the voltage to 11 000 for transmission 22 miles to Buffalo. There the voltage was stepped down to various voltages for a.c. lighting and power applications. It was also converted to d.c. for supplying an existing electric street railway system, as well as existing d.c. arc and incandescent lighting circuits.[124]

The Niagara Falls installation marked the beginning of the modern era of large-scale, hydro-electric plants transmitting polyphase a.c. over extended transmission and distribution systems. The feasibility of large-scale electrical power projects had been proven, opening the way for generating plants which would soon dwarf the Sault Ste. Marie Canal d.c. powerplant — and all other isolated and central stations on the d.c. system — into insignificance. Prior to 1893, 100- to 200-horsepower dynamos and 500-horsepower turbines were among the largest in operation anywhere, and no transformer had been built with a capacity of over 10 horsepower. But the mammoth Niagara power system proved highly efficient and flexible. On completion, it became the model for electrical power plants in North America.[125] Subsequently, all new hydro-electric power and lighting plants were on the polyphase a.c. system.[126] Within a year, it was noted that 2000-horsepower dynamos were common in new hydro-electric power installations.[127]

As of 1897-98, when consideration was being given to converting the d.c. electrical power and/or lighting installation on the 'Soo' Canal to a.c., the new polyphase a.c. power and lighting system was becoming well-established in Canada. Plans were already well advanced for installing a polyphase a.c. power and lighting system on the new Soulanges Canal, and several major hydro-electric power plants had already been completed on that system in Quebec: at the St. Narcisse waterfall on the Batiscan River ca. 1895; at the Montmorency Falls ca. 1896; at the Lachine Rapids on the St. Lawrence River in August 1897; and at Chambly on the Richelieu River in September 1897. Moreover in Ontario, a major hydro-electric power plant was nearing completion at the De Cew Falls on the Niagara Escarpment not far from the Welland Canal.[128]

Where canals were concerned, with the polyphase a.c. system it was possible to completely illuminate any waterway for efficient nighttime operation regardless of its length. As well, it was practicable for the first time to distribute power efficiently to any point along the navigation from a single powerhouse. The ready availability of water power on site, however, negated any need for the remote large-scale hydro-generating plants that came to be associated with the long-distance transmission lines and extensive, highly dispersed, distribution areas of the new polyphase a.c. era. Consequently, canal powerhouses erected on the new a.c. system remained on a comparatively small scale — only somewhat larger than the older d.c. hydro-electric stations such as on the Sault Ste. Marie Canal.

On the Soulanges Canal, which opened in October 1899, a single powerplant and generator complex operated a universal polyphase a.c. system for distributing both power and lighting along its whole 14-mile length. This system proved highly efficient, and the Soulanges installation became the prototype for the subsequent electrification of Canadian canals on the Great Lakes-St. Lawrence navigation system: on the eight-mile-long Lachine Canal, and the 12-mile-long Cornwall Canal in 1901-02; and the 27-mile-long Welland Canal in 1904-07.[129]

A 20-foot head of water was used on the Soulanges Canal to drive eight Victor turbines direct-connected to two 264-kW dynamos (and belt-connected to two exciters of 15 kW each) that produced three-phase, 60-cycle, alternating-current for transmission on a parallel primary circuit at 2500 volts. At each of the five 280 foot by 46 foot locks on the 14-mile-long canal, there were two transformers of 7500 watts capacity in a switch cabin. The transformers stepped down the voltage to 200 volts to operate eight 15-horsepower a.c. motors — one for each gate leaf and sluice culvert valve — on a secondary parallel circuit. Motors for working the waste weirs and swing bridges were also worked off these secondary circuits, as well as the incandescent lighting and electric heating units in the canal buildings.

The canal arc lighting ran off the primary circuit. A total of 220 long-burning, alternating-current enclosed arc lamps of 2000 candlepower were placed on 30-foot-high cedar poles at 480-foot intervals along the canal, and at 120-foot intervals beside the locks. A transformer of 1000 watts capacity stepped down the voltage at each lamp.

In the subsequent canal installations only minor changes were made. The arc lamps were grouped on secondary circuits of 25 lamps to enable a single transformer to serve a group of lamps, and in the case of the Cornwall and Welland canals, electricity was purchased from new polyphase a.c. power plants constructed nearby by private companies.[130]

After the Sault Ste. Marie experience, major ship canals constructed elsewhere in the world — such as the 51-mile-long Panama Canal built in 1904-15 — were electrically powered and illuminated. These canals were almost invariably operated on the polyphases a.c. system.[131] Indeed, the distribution limits inherent in the d.c. system made electrification impractical for canal systems of any great length with locks dispersed throughout. Engineers, however, did not totally reject the d.c. system. It was efficient and reliable within its limits as proven on the Sault Ste. Marie Canal.[132]

Operation and Maintenance

From its official opening in September 1895, the 900 foot by 60 foot lock on the new canal at Sault Ste. Marie lived up to all expectations, operating around the clock seven days a week. In conjunction with the two parallel locks on the American St. Mary's Falls Ship Canal — the Weitzel, and the new Poe lock opened in August 1896 — the Canadian 'Soo' lock was thoroughly tested on what was by far the world's busiest ship canal system. In its first year of full operation, the Canadian canal locked through over 400 vessels, both passenger steamers and freighters, carrying almost 400 000 tons of freight.[133] Vessels, which previously had to wait anchored in line anywhere from 12 to 36 hours to lock through the American canal, now passed with little or no delay through either of the two canals in operation.[134]

For the better part of two decades, 1895-1913, the Canadian lock was second to none in its efficiency of operation as shipping traffic boomed to the point where for several years the Canadian canal alone was the busiest in the world. In succeeding decades, the Sault Ste. Marie Canal ceased to rank in the forefront of either world or Canadian ship canals in freight tonnage. Nonetheless it continued to function as a major commercial navigation for more than 80 years: under the Department of Railways and Canals (1895-1936); the Department of Transport (1936-59); and the St. Lawrence Seaway Authority (1959-79).

The Growth Years: (1895-1913)

Freight traffic at Sault Ste. Marie consisted for the most part of iron ore and grain moving eastward from Lake Superior, and coal moving westward. Iron ore constituted by far the major tonnage as shipments to eastern foundries continued to surpass even the phenomenal rate of growth of the late 19th century. From 10 million tons shipped through both the Canadian and American 'Soo' canals in 1895, iron ore shipments reached 41 million tons by 1907.[135] Ore shipments accounted for about 67 percent of freight traffic with coal and grain, both increasing equally rapidly in volume, constituting 17 and 11 percent respectively.[136]

During the 1890s, one of the major objectives of the National Policy had been achieved with the spread of western settlement and the development of a wheat economy based on exports. In two decades with a heavy immigration of settlers onto the Canadian prairies, the acreage devoted to grain crops jumped from 1 209 338 to 16 853 000 acres. Grain exports boomed. Canada quickly became the world's third greatest wheat exporting nation, and a country widely acclaimed as "the granary of the British Empire" and future "Bread Basket of the World." Canadian wheat exports alone grew from 422 274 bushels in 1890, with year to year fluctuations, to 16 844 650 bushels in 1900.[137] Five years later over 68 million bushels of wheat were being shipped through both 'Soo' canals each navigation season. In 1906 this total soared to 84 million bushels, and dramatic gains were realized for all grain shipments — wheat, barley and oats — in most subsequent years.[138]

Overall, the total freight tonnage through the two canals at Sault Ste. Marie soared: from 15.22 million tons in 1895 to 62.4 million tons in 1910.[139] It far outstripped all other ship canals in the world. At the turn of the century, the annual tonnage was double the Suez Canal total, eight times that of the Kaiser Wilhelm (Kiel) Canal, and well over 10 times the tonnage on the Manchester Ship Canal. Moreover, both 'Soo' canals had only a seven- to eight-month navigation season, late April to early December; whereas the other canals were open year round.[140]

The Canadian 'Soo' Canal carried a substantial, rapidly growing annual tonnage during its first decade of operation: 2.04 million tons in 1900 increasing to 5.47 million tons in 1910. But this accounted for only eight to 14 percent of the phenomenal tonnage passing through the dual locks on the American canal in any given year. Thereafter, however, the largest of the ore carriers began to switch to the Canadian canal to maximize their carrying capacity. Consequently, in the period 1911-12 almost 55 percent of the total freight tonnage passed through on the Canadian side. In 1913, this reached 58 percent with the Canadian canal carrying an all-time high of 42.7 million tons of freight.

Within Canada, the 'Soo' Canal far outdistanced all other canals in freight tonnage. In 1896, it carried 57 percent of the total Canadian canal tonnage — over four times that of either the Welland or the St. Lawrence canals, its nearest competitors. In the 1913 record year, the 'Soo' Canal accounted for 82 percent of Canada's total canal freight, exceeding 10 times the tonnage on either the Welland or the St. Lawrence canals.[141]

At Sault Ste. Marie, there were no tariffs and/or transit fees, ships simply chose whichever canal promised the faster passage. Generally the smaller freighters, including the 255 foot by 44 foot Welland Canallers built for the Canadian grain trade, favoured the Canadian canal which could pass three at a lockage quickly in line. For some time, passenger steamers divided almost equally between the two canals, with each canal passing about 25 000 passengers a year.[142] The huge upper lakes freighters, however, were the primary carriers of iron ore and coal, as well as both American and Canadian grain. The routing of these vessels followed a changing pattern.

When the Canadian canal opened most ore carriers were 320 feet in length with a 40- to 50-foot beam with several up to 400 feet by 43 feet — vessels as large as many sea-going ships.[143] All such carriers, regardless of size, were able to pass both the American Weitzel lock and the Canadian lock two at a lockage, and the new American Poe lock was able to pass four of the 320-footers or two of the 400-footers at a lockage. In effect, the American canal enjoyed an almost three to one superiority in locking capacity where the huge ore carriers were concerned.

This advantage was maintained for a decade as ore carriers increased in size, but was progressively negated after 1906 when shipbuilders constructed ore carriers to a new 600 foot by 58 foot standard — too large for the Weitzel lock.[144]

These '600-footers' could pass through either the Poe or the Canadian lock, one vessel per lockage, but came to prefer the greater draught of the Canadian canal. Previously shallows in the river channels had limited ship draughts to 16 feet, and after 1898 to 18 feet, but as of 1906 the rivers and major upper lakes harbours had been deepened to 22 feet providing a 20-foot draught throughout.[145] The minimum draught in the upper lakes navigation system was henceforth at the 'Soo' canals and here the Canadian lock, with 20 feet three inches of water on the sills at extreme low water, had an almost eight-inch advantage over the Poe lock.[146] For the iron ore carriers, every inch of additional draught at that depth meant another 150 tons of carrying capacity.[147] Moreover, an overloaded vessel did not have to fear inordinate delays at the Canadian lock.

Vessels required an absolute minimum of a foot of water, and more generally two feet, under them to operate their propellers without settling on the bottom. An overloaded vessel could pass through the navigation system without difficulty, but not at the locks. On the Poe lock tugs had to be employed to tow any overloaded vessel out of the lock chamber, causing delays of up to an hour. This was not the case in the Canadian lock where vessels with only six inches of water under their bottom were passed. Taking advantage of the precise control afforded by the electrical system of operation, the filling valves were simply opened slightly to flash or flood the overloaded vessel out of the lock chamber.[148]

With the dramatic increase in shipping from year to year, traffic congestion became a serious problem again by 1910. At times anywhere from 10 to 40 vessels were at anchor waiting to pass through the Canadian lock alone, and on the busiest days as many as 56 vessels were put through in some 34 lockages. On several occasions when the Poe lock was closed for repairs, the traffic congestion was all but overwhelming. In one instance, the Canadian lock was in continuous operation for a 264-hour period during which 460 vessels of 1.4 million registered tons were locked through. As many as 87 vessels were anchored in the river above the lock awaiting passage, and 25 vessels below, with delays as long as 60 to 70 hours being experienced.[149]

Despite heavy usage, round-the-clock, relatively few problems were experienced in operating the Canadian lock during these years. Such problems as there were, with the lock gates and machinery in the powerhouse, were overcome during the off-season, avoiding any interruption to shipping.

After only two years of operation, the bowstring truss lock gates had shown signs of weakness, necessitating their being strengthened. Ultimately in 1900-01, the lower and intermediate gates were replaced with solid timber gates, built of Douglas fir beams framed with steel I-beams, and in 1910 the shorter upper gates and the guard gates were similarly replaced (Figs. 51 and 52).[150] During this period, it also proved necessary to set the sheaves of the gate opening cable system down into the floor of the lock chamber to prevent their being torn away by overloaded vessels scraping against them (Fig. 52).[151]

As early as 1906, the d.c. arc lighting system was converted to the a.c. system. In the powerhouse, a new 75-horsepower, a.c. arc lighting dynamo replaced the two d.c. arc lighting dynamos: viz. the No. 8 Wood, 9.6-amperes dynamo installed in 1898; and, its back-up, the original No. 7 Wood, 9.5-amperes dynamo. A new ammeter, volt meter, and circuit breaker, were installed on the switchboard, and transformers, lightening arresters, and new arc lamps placed in the outdoor lighting system along the canal (Fig. 54).[152]

In the canal powerhouse, a continuing problem was experienced with the pumping system for dewatering the lock chamber. The vertical shafts of the two pumps proved too light, necessitating their replacement. Coupling the two turbines to run the dewatering pumps also proved difficult, if not impossible, as the south turbine ran much faster than its twin. Efforts to govern the speed of the turbine failed. Finally in 1910-11, the intake and discharge pipes were modified and a new more powerful turbine installed with a governor (Fig. 53).[153] This 210-horsepower turbine was strong enough to drive the dewatering pumps by itself, as well as either of the two direct current dynamos of the electrical power system.[154] The north turbine was retained for driving the arc lighting dynamo, and as a backup for running the electrical power generators.[155]

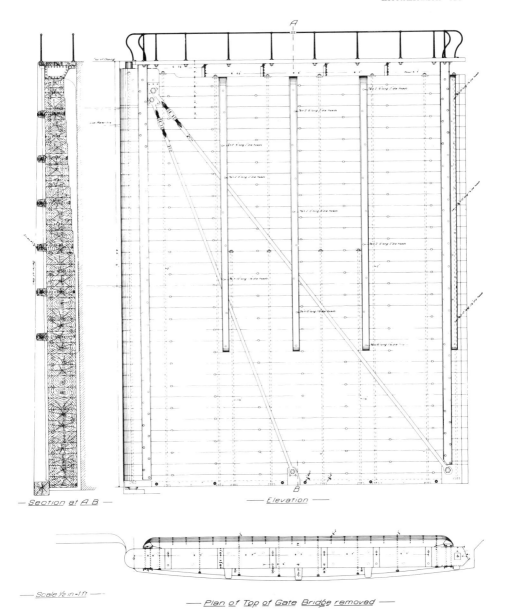

— Section at A B —

— Elevation —

— Scale ½ in - 1 ft —

— Plan of Top of Gate Bridge removed —

FIGURE 51. Drawing of solid timber lock gate, introduced in 1900-1901. ("Plan of Lower Main Gates," 26 September 1900, Canadian Parks Service, Sault Ste. Marie Canal Office)

FIGURE 52. The modified gate cable system, with snub pulleys recessed in the mitre sill and at the base of the lock wall to enable a single screw power machine to both open and close a gate leaf. (Canadian Parks Service, Sault Ste. Marie Canal Office)

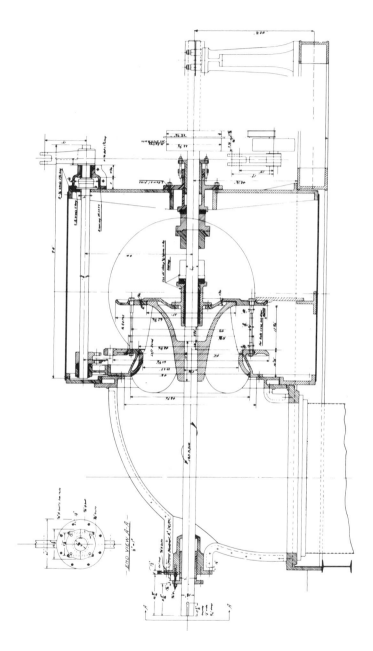

FIGURE 53. Drawing of new 210-horsepower turbine installed in 1910-11. ("General Drawing, Power House Turbine," n.d., Canadian Parks Service, Sault Ste. Marie Canal Office)

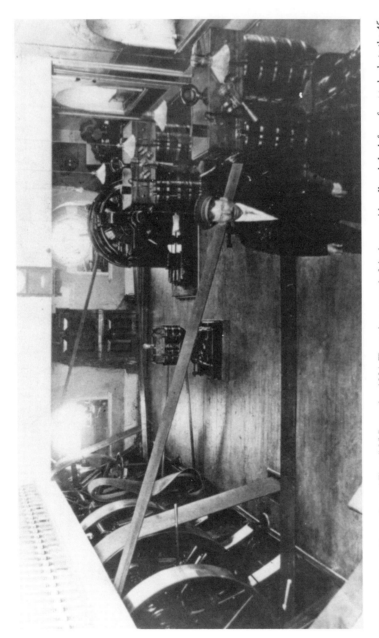

FIGURE 54. "Power House, Second Floor," 25 September 1934. The main countershaft is shown with pulleys belted, from front to back, to the 45-kW power dynamo, the south turbine below, the backup 45-kW power dynamo (belt disconnected), the 75-kW arc lighting dynamo, and the north turbine below. (Canadian Parks Service, Sault Ste. Marie Canal Office)

The incandescent lighting system continued to be run off the original three-kW dynamo, driven by the 13-inch auxiliary turbine, until 1921. In that year, a new six-kW, 110-volt dynamo was installed in the powerhouse, with new 500-watt lamps on the indoor incandescent lighting circuit.[156] No problems were experienced in driving the larger dynamo off the auxiliary turbine.

During its boom years, 1911-13, the Canadian lock was as efficient as any lock in the world, if not more so (Fig. 55). This advantage, however, was soon lost as the American government completed the first of what would be a series of new locks on ever greater scales of construction to meet the unprecedented shipping demands at Sault Ste. Marie, Michigan.

FIGURE 55. A whaleback lake boat leaving the Sault Ste. Marie Ship Canal lock. (K. Elder, postcard collection)

American Dominance: (1914-79)

By 1907, the first of a new generation of ore carriers, vessels 625 feet by 60 feet and larger, had been launched on the upper lakes as shipbuilders strove to increase shipping capacity. These vessels were too wide for the Canadian lock, but could pass through the Poe lock one at a time (Fig. 56).[157] The American government had responded by undertaking the construction of a new canal with two parallel locks: the Davis and the Sabin. The new locks were based on the Canadian

FIGURE 56. The *S.S. Lemoyne* on Fourth Welland Canal, during its official opening 6 August 1932. Launched at Midland, Ontario, in 1926, it heralded a new generation of lake boats too large for the Sault Ste. Marie Ship Canal. (National Archives of Canada, RD-989)

long-lock configuration and electrically powered, with electric lighting along the whole canal for round-the-clock operation.

The new locks were as efficient in operation as the Canadian lock, but far larger with a much greater lockage capacity. Each of the 1350 foot by 80 foot concrete locks had 24 feet six inches of water on the sills, and was designed to pass two of the "625-footers" at a lockage. The new ship locks were on a scale in keeping with the largest locks for ocean-going vessels, namely the six pairs of 1000 foot by 110 foot concrete locks on the Panama Canal. On its opening in August 1914, the Panama locks had deprived the Canadian 'Soo' lock of its distinction as the longest ship canal lock in the world prior to its again being superseded by the Davis and Sabin locks. When opened in late October 1914 with the Davis lock in operation and the river channels dredged to provide a minimum 21-foot draught, the new system had a devastating impact on the Canadian canal.[158]

Tonnage through the Canadian lock dropped dramatically from the all-time record high of 42.7 million tons in 1913 to 7.8 million tons in 1916. With the opening of the Sabin lock in 1919, freight dropped off again to about two million tons where it remained over subsequent decades. In contrast, traffic on the American canals continued to increase from 63.5 million tons in 1915, with wide fluctuations, to 87.9 million tons in 1940 — more than sufficient to maintain the 'Soo' canals system as by far the busiest in the world. Only in passenger traffic did the Canadian canal continue to prosper. While the American canals experienced a severe decline, passenger traffic on the Canadian canal jumped up over 30 000 a year, and with the exception of the depression years, continued to grow. Passenger ships, in effect, sought to avoid the heavy freight traffic on the American canals.

As freight tonnage plunged, the Sault Ste. Marie Canal lost its predominant position in the Canadian canal system. On the Welland and St. Lawrence canals tonnage continued to increase until by the mid-1920s both canal systems were carrying four times more than the Canadian 'Soo' canal.[159] The increase on the other canals was due in large part to the continued growth of the Canadian wheat export trade. By 1923, Canada surpassed the United States as the world's leading wheat exporter,[160] and by 1930-31 would account for 40 percent of the world's wheat trade, exporting some 258 million bushels.[161] Although constructed to serve as the last link in the Canadian wheat transportation system, the Canadian canal at Sault Ste. Marie benefited little from the phenomenal growth of wheat exports. Over 95 percent of the Canadian wheat shipped via the 'Soo' passed through the new American canal.[162]

During the Second World War as the North American steel industry boomed, the American canals attained the first of a succession of 100-million-ton years.[163] Freight traffic on the Canadian canal, however, continued to fluctuate around two million tons a year with two brief upsurges. It reached 4.5 million tons in 1943; and 3.38 million tons in 1953. But in the best years the Canadian canal carried only 2.5 percent of the total tonnage passing through the 'Soo', and the gains were only temporary aberrations.[164] The war-time surge was curtailed in July 1943 when a new 800 foot by 80 foot lock, the MacArthur lock with 31 feet of water on the sills, replaced the obsolete Weitzel lock.[165] This equipped the American system with four locks larger than on the Canadian canal, three of which provided a much greater depth of draught. These parallel locks were more than adequate to meet all shipping demands as tonnage at the 'Soo' increased to an all-time record of 128.5 million tons in 1953 — a high which would never be attained again owing to a rapid decline in coal shipments.[166]

In subsequent decades as ore carriers continued to increase in size, it was the American canal system that was again expanded to meet this new demand. During the 1960s, the 1896 Poe lock was replaced with a new 1200 foot by 110 foot Poe lock with 32 feet of water on the sills.[167] With four comparatively large and deep draught locks in operation on the American dual canal system — the Davis and Sabin; and the MacArthur and new Poe lock — freight traffic on the Canadian canal continued to decline reaching less than 0.5 million tons per annum by the mid-1970s. In contrast, passenger traffic continued to increase with as many as 195 000 passengers passing through the Canadian canal in any given year, on both tour and pleasure boats.[168]

In recognition of this trend, the St. Lawrence Seaway Authority began to prepare plans for phasing the Sault Ste. Marie Canal out of commercial operation.[169] Ultimately, in April 1979, it was turned over to Parks Canada (now the Canadian Parks Service) for preservation as a heritage canal.[170] By that time, however, a number of changes had been made in the appearance and operating equipment of the Sault Ste. Marie Canal. Most of these date from the 1960s, with but one major exception — the conversion of the electrical power system for operating the lock.

Alterations (1930-85)

For nearly 50 years after its opening in 1895, the electrical power system and lock operating machinery performed flawlessly, with only routine maintenance being required.[171] Then in response to increasingly heavy war-time traffic, changes were made. At the close of the 1942 navigation season, the electrical power system was converted to a.c., abandoning the d.c. dynamos in the canal powerhouse. Six new 10-horsepower, wound rotor motors were installed to operate the lock gate screw machines, and new 15-horsepower, squirrel-cage motors on the four valve screw machines. The wiring system was totally renewed, and the old arc lighting system replaced with more modern incandescent outdoor lamps.[172] The new lamps remained in service for almost three decades and were replaced in turn, in 1968, with a new type of mercury arc lamp. At the same time, cedar poles gave way to concrete standards around the lock and steel standards on the piers and approach walls.[173]

With a.c. electricity being purchased nearby from the Great Lakes Power Company, the two main turbines in the powerhouse were used after 1942 only for driving the centrifugal pumps when dewatering the lock chamber. As late as 1963, both turbines remained operational,[174] but thereafter the north turbine — the last of the two original 155-horsepower "American turbines" of 1895 — was removed. The larger 210-horsepower turbine installed in 1910-11, was retained for driving the two dewatering pumps (Figs. 57 and 58).[175]

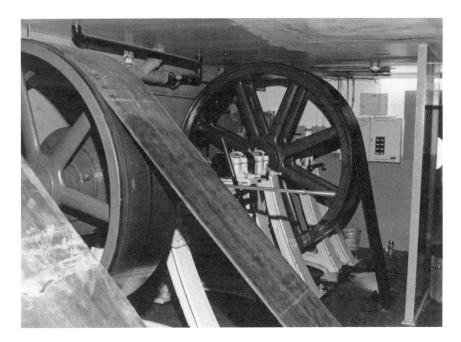

FIGURE 57. Countershaft on second floor of powerhouse, showing two belts driving the dewatering pumps system below, and the drive belt from the remaining turbine below. (R. Draycott, Sault Ste. Marie Canal Office, 10 March 1986)

With the exception of the dewatering pumps and their turbine driven gearing system, all of the original powerhouse equipment was removed with the north turbine. This apparently took place during the winter of 1963-64 when new support columns and a concrete floor were constructed in the powerhouse.[176] Only two artifacts were saved from the powerhouse electrical equipment: one of the 45-kW, 500-volt, Edison bi-polar dynamos; and the electrical switchboard. They were given to the Ontario Hydro Museum in Toronto, and subsequently loaned to the National Museum of Science and Technology in Ottawa (Fig. 44 and 59).[177]

At the lock, the original gate and sluice valve screw machines continued to operate as designed until 1967 when the wiring system was modified to interlock the gate and valve machinery in a single cycle of operation.[178] Otherwise, only routine maintenance was performed, consisting for the most part of replacing the gate opening cables each year, the gate closing cables every three years, and the bronze bushing and screw shafts at 10-year interval.[179] But heavy wear on the screw machines of the lower main gates, machines No. 1 and No. 2, finally necessitated their replacement. During the winter of 1971-72, two modern Jactuator machines were installed to operate the lower main gates.[180] The other four gate

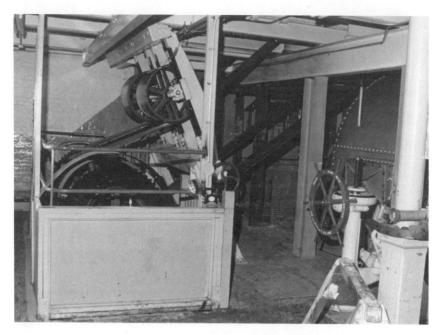

FIGURE 58. First floor of the powerhouse, showing bevelled-gear drive system for the dewatering pumps below in the pump well. The two belts are deiven from the countershaft on the second floor. At right: scroll case of the 1910-11 turbine and control wheel of the turbine governor. (R. Draycott, Sault Ste. Marie Canal Office, 10 March 1986)

screw machines — on the upper ages and the auxiliary gates — have remained untouched (Fig. 60) as have the four valve-operating screw machines.

In the lock chamber, the original machinery continues to operate, but the lock walls have been transformed in appearance. In 1933, a serious leak developed through the south lock wall at the upper main gates, resulting in large quantities of face stone being forced out. To seal the leak a concrete cut-off wall was built out for a distance of 100 feet perpendicular to the lock, and a large area of the south wall cut back and faced with concrete. Cracks in the remaining stone masonry facing were grouted to prevent further seepage.[181] Subsequently, the remainder of the south wall was also faced in concrete. In response to serious leakage through the north wall in 1959-62, the whole stone masonry face was cut back anywhere from 12 inches to 32 inches and rebuilt with reinforced concrete.[182]

Surprisingly few problems have been experienced with the timber and plank floor culverts of the Sault Ste Marie Canal lock over the years. The first serious problem involved the discharge culverts. In May 1946, after 51 years of operation, a 1200-square-foot area of flooring collapsed down into the discharge culverts at the lower main gates. The canal had to be closed for 43 days to effect repairs.[183]

FIGURE 59. Edison bi-polar dynamo on the second floor of the powerhouse with switchboard at left, 25 September 1934. Both the switchboard and one of the Edison dynamos have been preserved and are in storage at the National Museum of Science and Technology in Ottawa. (Canadian Parks Service, Sault Ste. Marie Canal Office)

FIGURE 60. Horizontal screw power machine, showing drive gears and travelling cross-head mounted on two threaded shafts. (Canadian Parks Service, Sault Ste. Marie Canal Office)

FIGURE 61. Discharge valves, shortly after the timber discharge culverts were rebuilt in reinforced concrete during the winter of 1963-64. (Canadian Parks Service, Sault Ste. Marie Canal Office)

The discharge culverts functioned well for another 17 years, but then a severe leakage problem developed below the lower gates. In response, the discharge culverts, and the lower and auxiliary gate sills, were totally reconstructed in reinforced concrete during the winter of 1963-64 (Fig. 61).[184] At the same time, new bearings were installed on the horizontal shafts of the butterfly valve plates in the newly reconstructed culverts.[185]

Other than replacing the bearings, the four discharge valves and the four butterfly valves on the filling culverts, have required very little maintenance to the present. The same is true of the timber filling culverts and plank floor in the lock chamber.

Where the filling culverts are concerned, the only interruption of shipping occurred in late September 1961 when a section of floor near the upper gates blew up. The canal was closed for 43 hours while repairs were made. The following winter, the timber culvert walls and a new plank floor — consisting of a new two-inch plank covering and a partial renewal of the three-inch plank subfloor — were stiffened and tied down with long bolts anchored into the bedrock below.[186] A similar system had been used to strengthen a section of floor in 1943, and would be used on an even more extensive scale in 1971 when sections of the plank floor again showed signs of heaving.[187] Otherwise, the timber filling culverts and the plank flooring in the lock chamber have remained as originally constructed.

In recent years, leakage has again become a serious problem in the 'Soo' Canal lock. The two original centrifugal pumps were designed, with a combined pumping capacity of 32 000 gallons per minute, to dewater the lock chamber in six to seven hours.[188] But by 1963, it took 12 hours to dewater the lock and when completely dewatered the pumps still had to be kept working at full capacity to keep up with the leakage.[189] Subsequently auxiliary pumps were installed in a temporary pumphouse to increase pumping capacity, but the situation continued to worsen. By 1984, it was taking up to 16 hours to dewater the lock chamber with the leakage volume calculated at between 6000 and 8000 U.S. gallons per minute.

During the winter of 1984-85, the temporary pumphouse was removed and three 50-horsepower vertical turbine pumps, manufactured by Worthington Canada Ltd., were installed in the canal powerhouse for dewatering the lock. The two original centrifugal pumps, however, were retained as back-ups, owing in large part to a conscious effort by Parks Canada to preserve the historic pumping machinery.[190] The centrifugal pumps are still turbine powered in the same manner as in 1895. The belting and gears on the horizontal and vertical shafting of the drive system, however, have been renewed in kind.[191]

A World's First

The Sault Ste. Marie Canal was the first ship canal in the world to be electrically operated, and the first to be totally electrified with electrical power and lighting along the entire canal. The decision to employ electrical power to overcome a potential cold-weather operating problem with the traditional hydraulic system of operation, resulted in a highly innovative engineering undertaking. It involved a major and totally novel application of electrical power, but one that made use of an existing technology well proven during the previous five to eight years on electric street railway systems.

In consultation with C.G.E., the canal engineers were able to design and install a highly efficient and self-contained electrical power system that performed well under all weather conditions. Indeed, the system yielded some 47 years of almost trouble free operation until converted to polyphase a.c. power in 1942-43. The 'Soo' Canal installation incorporated many of the latest technical developments in d.c. power, as well as d.c. arc and incandescent lighting systems. It was, however, only one of many new electrical power applications in a period, 1893-

95, when North American engineers were actively developing and consciously seeking new uses for electrical power. The screw machines designed to enable electric motors to operate the lock gates and floor sluice valves were ingenious, but no doubt based on contemporary machine tool design technology.

Short sections of several North American canals had been illuminated electrically prior to the construction of the Sault Ste. Marie Canal. But distribution limits inherent in the d.c. system had precluded a more extensive system of electric lighting. Only the exceptionally short length of the 'Soo' Canal — about one mile — enabled all the buildings, the lock, and the navigation channel to be fully illuminated with electricity. Once again, however, the lighting technology employed had been well established for over a decade: the incandescent system for office and factory lighting; and the arc lamp system for street lighting. Nonetheless, it was the lighting systems that proved the least satisfactory, resulting in the arc lighting system being converted to a.c. as early as 1906, and the incandescent lighting in 1921.

The Sault Ste. Marie Canal powerhouse in its scale of construction, layout and electrical equipment was a typical low-head, d.c. hydro-electric plant of the late 1880s and early 1890s. Yet in many ways, it represents a transition phase in electrical power generation and distribution. It was erected in a watershed period, 1893-95, when d.c. systems of electrical power and lighting were on the verge of being rendered all but obsolete because of their inherent limits of scale and transmission capabilities in the face of on-going developments in the newer polyphase a.c. technology.

The characteristically small d.c. power plant providing electrical power and/or lighting to an adjacent distribution area of several square miles in extent, or to a limited area along one section of a canal, soon gave way to huge polyphase a.c. hydro-electric plants transmitting electricity at unprecedently high voltages over widely extended transmission and distribution systems employing large transformers and transformer substations. At the same time, horizontal shaft turbines and dynamos of less than 200 horsepower, gave way to huge direct-connected vertical shaft turbines and dynamos rated at thousands and soon tens of thousands of horsepower. This precision engineered equipment, designed for specific installations, put an end to the older system employing off-the-shelf turbines and dynamos made to work properly on site through the "cut and try" method and adjustments in the pulley ratios of the belt-drive systems. These new polyphase a.c., hydroelectric power plants have dwarfed the 'Soo' Canal powerhouse into insignificance, but by no means diminished its historic significance.

The electrification of the 'Soo' Canal marked the introduction of electrical power for operating large ship canal locks around the world. It proved the feasibility and efficiency of electrical power in operating ship locks, and inaugurated a new era where ship canal locks were concerned. It did not, however, establish the technology employed thereafter.

Where new ship canals were constructed, or existing canals electrified, it was almost invariably done on the polyphase a.c. system. No matter how extensive the canal, on the universal polyphase a.c. system electricity for both power and lighting applications could be provided from a single generating plant and distribution system through employing transformers on a parallel circuit. In many instances, electricity was simply purchased from existing private polyphase a.c. hydro-electric systems, dispensing with the need for a canal powerhouse. Indeed, the Sault Ste. Marie Canal was the only major ship canal ever to be totally electrified on the d.c. system.

The 'Soo' Canal powerhouse remains as originally constructed, but the interior has been greatly modified and all of the electrical equipment removed. One turbine remains — the 210-horsepower turbine installed in 1910-11 — and is used exclusively for operating the two original dewatering pumps in a backup capacity (Figs. 57 and 58). This horizontal shaft turbine and the pump driving machinery — the belting, countershafts, and gear shafts — provide a graphic view of the layout and drive system formerly employed to power the dynamos of the electrical power system and the two lighting systems.

The lock is still operated in the same manner as in 1895 with four out of the six original lock gate screw power machines remaining, as well as all eight original steel butterfly valves, and the four screw power machines and draw-rod/crank systems for working the valves. They, however, are now driven by modern a.c. motors deriving their current from the local power grid. The floor culverts system continues to function as originally designed, but the discharge culverts have been rebuilt in reinforced concrete. The plank floor, and timber filling culverts remain as originally built. The appearance of the lock has been altered somewhat, through the stone masonry walls being refaced in concrete, but otherwise the lock remains as constructed. The lock gates are replicas of the solid timber gates introduced in 1900-01.

Today, the 900 foot by 60 foot Canadian 'Soo' lock is the sole survivor of the several colossal structures constructed at Sault Ste. Marie in the late 19th century. Together with the now demolished Weitzel and original Poe locks on the St. Mary's Falls Ship Canal, the Sault Ste. Marie lock set a new standard of scale for large ship canal locks. Since its opening in September 1895, the electrically powered lock has provided over 90 years of relatively trouble free service. Having long since passed its heyday as one of the world's busiest commercial navigations — and, for a brief time, the world's busiest ship canal — it is now operated by the Canadian Parks Service as a heritage canal for the benefit of pleasure craft and tour boats.

Endnotes

1   Sault Ste. Marie [SSM] Canal Office, Drawing No. 490-1, Case D - 155, "Sault Ste. Marie Canal, Plan of Lock as Enlarged," T. Trudeau, Chief Engineer of Canals, 5 April 1892. See also Thomas Keefer, "The Canals of Canada," Plate III.

    On the hydraulic system of the Weitzel lock see: E.S. Wheeler, "Locks of the Nicaragua Canal and St. Mary's Falls Canal," *Engineering News and American Railway Journal* [hereafter cited as *Engineering News*] (1 June 1893), p. 505.

2   Canada, Department of Railways and Canals [hereafter cited as DRC], *Annual Report*, 1893, Appendix No. 7, pp. 125-126.

3   "Power Transmission," *Electrical Engineer* (16 October 1895), p. 380.

4   DRC, *Annual Report*, 1893, pp. 126 and 129.

5   Ibid., 1891, pp. 101-102. The Edison electric plant had been transferred to Beauharnois from the Lachine Canal. There a new arc lighting system replaced the former incandescent lighting system along the canal.

6   Ibid., p. 125; and "Operation of Locks by Electrical Power - The new Experiments on the Beauharnhois Canal," *The Canadian Engineer*, Vol. 1 (August 1893), p. 91.

7   "The Sault Ste. Marie Canal," *Scientific American Supplement* (6 September 1890), p. 12231.

8   "Power Transmission," *Electrical Engineer* (16 October 1895), p. 381; and "The Canadian Ship Canal at Sault Ste. Marie," *Engineering and Building Record* and *Sanitary Engineer*, [hereafter cited as *Engineering Record*], Vol. 32, No. 25 (16 November 1895), p. 436. The hydraulic cylinders for operating the filling and discharge valves were under water at the bottom of the lock, and hence not subjected to freezing temperatures.

9   D.A. MacGibbon, *The Canadian Grain Trade* (Toronto: Macmillan of Canada, 1932).

10 "The Canadian Ship Canal Lock at Sault Ste. Marie, Ont.," *Engineering News* (28 March 1895), p. 207.

11 DRC, *Annual Report*, 1894, Appendix No. 6, p. 122.

12 "Power Transmission," *Electrical Engineer* (16 October 1895), p. 381.

13 NA, RG43, B2e, Vol. 1772, Folio 476, "Specification of Water wheels, penstocks, inlet and discharge pipes and machinery for the powerhouse at Sault Ste. Marie Ontario Canal," 30 December 1893; and Canal Records, File C-4250/S32-1, Vol. 6, Report of Privy Council, 11 January 1894.

14   DRC, *Annual Report*, 1894, pp. 121-122; and NA, RG43, B2e, Vol. 1771, Folio 327, Tender for Electric Light and Power Plant, 29 March 1894.

15   See for example, Canal Records, File C-4250/S32-1, Vol. 6, Kennedy to Spence, "Extra Cost of Machinery and Belts arising from Changes required in Driving Centrifugal Pumps," 2 March 1894; and ibid., Report of Privy Council, 7 February 1894, wherein a change from 36-inch to 45-inch diameter turbines is authorized.

16   "The Canadian Ship Canal at Sault Ste. Marie," *Engineering Record* (16 November 1895), p. 436; and "Power Transmission," *Electrical Engineer* (16 October 1895), p. 380. See also NA, RG43, B2e, Vol. 1772, Folio 476, "Specification of Water wheels, penstocks, inlet and discharge pipes and machinery for the powerhouse...," 30 December 1893; and ibid., Vol. 1771, Folio 329, "Specification for the construction of Steel Pipes and the Valves," Sault Ste. Marie Canal, 13 July 1893.

17   Louis C. Hunter, *A History of Industrial Power in the United States, 1789-1930, Volume One: Waterpower in the Century of the Steam Engine* [hereafter cited as *A History of Industrial Power, Vol. 1*] (Charlottesville: University Press of Virginia/ Eleutherian Mills-Hagley Foundation, 1979), pp. 305-306.

      Turbines were introduced into the United States from France ca. 1842-44, and by the 1880s highly improved turbines had completely displaced older types of water wheels in American mills and industries (ibid.).

18   Ibid., p. 383 and 393-394. On the problems associated with step bearings see Arnold E. Roos, "Working Paper on Hydro-Electric Technology," unpublished manuscript (Ottawa: Environment Canada, Canadian Parks Service, 1985), pp. 90-91.

19   W.W. Tyler, "The Evolution of the American Type of Water Wheel," *Journal of the Western Society of Engineers* [hereafter cited as *J.W.S.E.*], Vol. 3, No. 2, (April 1898), pp. 893-897. In the early 20th century, the vertical shaft turbine supplanted in turn the horizontal arrangement, abetted by the development of the Kingsbury bearing, the vertical shaft umbrella dynamo, and lastly the limitations of size inherent in the horizontal turbine arrangement.

20   Roos, "Working Paper on Hydro-Electric Technology," pp. 56 and 106.

21   When first mounted on a horizontal shaft in 1879, the American turbine had proved up to 15 percent less efficient than in its traditional vertical shaft setting. The provision of a large draft tube was found to eliminate this comparative loss of efficiency (Tyler, "The Evolution of the American Type of Water Wheel," *J.W.S.E.* (April 1898), pp. 893-895.

22   The introduction and subsequent development of the 'American' turbine, as well as its working principle, are well described in Hunter, *A History of Industrial Power, Vol. I*, pp. 347-372. After 1915, the American turbine was superseded in turn by the revived Francis turbine which would predominate thereafter in major hydro-electric power station installations (ibid., pp. 391-392).

23   Tyler, op. cit., pp. 895-896.

24   Canal Records, File C-4250/S32-1, Vol. 9, letterhead, W. Kennedy & Sons to John Balderson, Secretary, DRC, 25 July 1895.

25   Hunter, *A History of Industrial Power, Vol. I*, p. 389.

In Canada, the Leffel turbine predominated followed closely by the Tyler turbine (Felicity L. Leung, "Direct Drive Waterpower in Canada: 1607-1910," Environment Canada, Canadian Parks Service, Ottawa, Microfiche Report Series, No. 271, 1986, pp. 59-62).

The Leffel turbine, developed by James Leffel of Springfield, Ohio, in 1852, was a unique and highly efficient double-wheel turbine combining a Francis type inward-flow upper runner with an inward and downward-flow lower runner (Hunter, *A History of Industrial Power Vol. I*, pp. 366-374).

The Tyler turbine, was a scroll turbine with a central discharge. Developed in the United States, and widely used after 1850, it was considered antequated by the 1890s (Tyler, "The Evolution of the American Type of Water Wheel," pp. 886-887).

26   Roos, "Working Paper on Hydro-Electric Technology," p. 50.

27   "The Canadian Ship Canal at Sault Ste. Marie," *Engineering Record* (16 November 1895), p. 436. Iron lamp standards are mentioned, but cedar poles were specified, and ultimately used, for the arc lamps (NA, RG43, B2e, Vol. 1771, Folio 327, Specification for Electric Light and Power Plant, 29 March 1894). Oak tanned leather belts of 11/16 inch in thickness, were considered the best belting for driving generators [W.J. Buckley, *Electric Lighting Plants, their Cost and Operation* (Chicago: William Johnston Printing Co., 1894), p. 124].

The enclosed arc lamp was developed in 1893. Open carbon arc lamps had an eight-hour life, but enclosed lamps lasted 100 hours or more. By 1895, the enclosed lamp was in common use in North America, bringing with it the phenomenon of all-night lighting, every night. Previously arc street lighting had been turned on only on dark nights, after midnight (Harold C. Passer, *The Electrical Manufacturers, 1875-1900, A Study in Competition, Entrepreneurship, Technical Change, and Economic Growth*, [Cambridge: Harvard University Press, 1953, reprinted 1978], pp. 64-65).

28 NA, RG43, B2e, Vol. 1772, Folio 476, Specification of Water wheels, penstocks, inlet and discharge pipes and machinery for the powerhouse, 30 December 1893.

29 "Power Transmission," *Electrical Engineer* (16 October 1895), p. 381. Originally, four propeller wheels were specified, but when tested proved too much of a load (NA, RG43, B2e, Vol. 1772, Folio 471, W. Kennedy & Sons to J.B. Spence, DRC, 9 September 1895 and 3 October 1895).

30 "The Canadian Ship Canal Lock at Sault Ste. Marie, Ont.," *Engineering News* (28 March 1895), p. 206. See also SSM Canal Office, Engineering Drawings, "Plan Showing the Position of the Discharge Pipes from the Large Pumps," J.B. Spence, 3 February 1894.

31 NA, RG43, B2e, Vol. 1772, Folio 471, W. Kennedy & Sons to J.B. Spence, 29 June 1894, Sketch of planned layout of Pulleys on main countershaft.

32 "The Canadian Ship Canal at Sault Ste. Marie," *Engineering Record* (16 November 1895), p. 436; and Canal Records, File C-4250/S32-1, Vol. 8, Kennedy & Sons to C. Schreiber, Chief Engineer, DRC, 29 January 1895.

33 "Power Transmission," *Electrical Engineer* (16 October 1895), p. 381; and NA, RG43, B2e, Vol. 1771, Folio 327, Tender for Electric Light and Power Plant, 29 March 1894.

34 "Power Transmission," *Electrical Engineer* (16 October 1895), p. 381.

35 Roos, "Working Paper on Hydro-Electric Technology," pp. 282-284. On the development of circuit breakers ca. 1880-95 see ibid., pp. 249-251, and for ammeters and voltmeters, ibid., p. 281.

36 DRC, *Annual Report*, 1894, Appendix No. 6, p. 122.

37 Canal Records, File C-4250/S32-1, Vol. 8, W. Kennedy & Sons to C. Schreiber, 29 January 1895.

38 DRC, *Annual Report*, 1896, Appendix No. 3, p. 97.

39 Canal Records, File C-4250/S32-1, Vol. 6, "Specification for Machinery for Operating Gates and Valves," 23 May 1894.

40 Ibid., Vol. 10, John J. McGee, Clerk, Report of Privy Council, 27 November 1896; and NA, RG43, B2e, Vol. 1772, Folio 471, F.J. Leigh, Superintendent, Canadian Locomotive and Engine Co., Kingston, to J.B. Spence, Designing Engineer, DRC, 11 August 1894. For example, the sliding pulley head frame had to be redesigned to avoid potential cracking problems in cooling the castings (ibid.).

41 DRC, *Annual Report*, 1895, p. lxxviii. For details on the components of the gate machine see Canal Records, File C-4250/S32-1, Vol. 6, "Specification for Machinery for operating Gates & Valves Collingwood Schreiber, Chief Engineer, DRC, 23 May 1894.

42   "The Canadian Ship Canal Lock at Sault Ste. Marie, Ont.," *Engineering News* (28 March 1895), p. 207 and Plates: Fig. 5, "Screw Power Machine Operating Gate Leaves"; and Fig. 5a, "Details of Gate Operating Machine"; and "Power Transmission," *Electrical Engineer* (16 October 1895), p. 380.

43   "The Canadian Ship Canal Lock at Sault Ste. Marie, Ont.," *Engineering News*, Fig. 6, "Plan Showing Attachment and Arrangement of Cables Operating Gate Leaves"; and "The Canadian Ship Canal Lock at Sault Ste. Marie," *Engineering Record* (16 November 1895), p. 436.

44   SSM Canal Office, Historic Prints Collection, negative A-22, "Connecting up Operating Cables," 12 June 1895.

45   "The Canadian Ship Canal Lock at Sault Ste. Marie, Ont.," *Engineering News* (28 March 1895), Fig. 4, "Screw Power Machine Operating Valves."

46   Ibid., p. 206; and "The Canadian Ship Canal at Sault Ste. Marie," *Engineering Record* (16 November 1895), p. 436. See also, NA, RG43, B2e, Vol. 1771, Folio 332, "Specification for Valves, including frames, shafts, pillow blocks, crank arms, draw-rods, securing bolts, grating, etc.," DRC, 30 November 1893; and Canal Records, File C-4272/S32, Vol. 2, "Sault Ste. Marie Canal, Part Details of Lock Showing Culverts etc. for Valve Shafts," T. Trudeau, Chief Engineer, DRC, 5 April 1892.

47   "The Canadian Ship Canal Lock at Sault Ste. Marie, Ont.," *Engineering News* (28 March 1895), p. 206; "The Canadian Ship Canal at Sault Ste. Marie," *Engineering Record* (16 November 1895), p. 436; and DRC, *Annual Report*, 1895, Appendix No. 7, p. 129.

48   Canal Records, File C-4250/S32-1, Vol. 9, Schreiber to J.H. Balderson, Secretary, DRC, 23 July 1895; and DRC, *Annual Report*, 1896, Appendix No. 3, p. 97.

49   DRC, *Annual Report*, 1895, p. lxxvii, and 1896, p. 97.

50   On the auxiliary buildings see Sally Coutts, "Sault Ste. Marie Canal Buildings, Sault Ste. Marie, Ontario," Building Report: 85-07, Federal Heritage Buildings Review Office, Environment Canada, Canadian Parks Service, 1985.
    See also NA, RG43, B2e, Vol. 1771, DRC, "Specification for six buildings for enclosing the several machines operated by electric motors for operating the valves and gates," 25 October 1894.

51   DRC, *Annual Report*, 1895, p. 49, "Capital Accounts - Canals," and ibid., 1896, p. 97. Subsequently following arbitration proceedings, Hugh Ryan and Co. were awarded an additional $211 505 for their work, plus six percent interest from 1 September 1894 to the date of the settlement (NA, RG43, Vol. 1698, File 4860, Report of Privy Council, 12 February 1900).

Ultimately, when all contracts were completed, the Sault Ste. Marie Canal cost $3 490 000, well below the $4 million revised estimate (DRC, *Annual Report*, 1896, p. lxx).

52   DRC, *Annual Report*, 1896, p. 98, 1897, p. 121, and 1898, p. 121; "The Canadian Ship Canal at Sault Ste. Marie," *Engineering Record* (16 November 1895), p. 436; and "Power Transmission," *Electrical Engineer* (16 October 1895), p. 380.

For the American locks see DRC, *Annual Report*, 1898, p. 121; Wheeler, "Locks of the Nicaragua Canal and the St. Mary's Falls Canal," *Engineering News* (1 June 1893), pp. 504-505; and Kibbee, "The Busiest Canal in the World," *Engineering Magazine*, Vol. 13 (July 1897), p. 610.

The Weitzel lock, with two floor culverts, took much longer to fill and empty — 12 minutes and eight minutes, respectively — but operated as quickly as the Poe lock because only two vessels, rather than four, had to be positioned in the lock chamber.

Within a decade, the greatly increased size of ore carriers would present a real danger of their surging into the lock gates on the water being let into the lock chamber too quickly. To avoid this threat, the operation was slowed somewhat. As a result, up lockages thereafter took about 20 minutes — still a significant advantage over the American locks (J.W. Le B. Ross, "General Design of a Lock and Approaches," *Journal of the Engineering Institute of Canada*, Vol. 3 [August 1920], p. 384).

53   DRC, *Annual Report*, 1897, pp. 119 and 121.

54   On the early development of the generator see Sigvard Strandh, *A History of the Machine* (New York: A. & W. Publishers, 1979), pp. 153-159.

The earlier magneto-electric generators, produced during the 1840s and 1850s primarily for lighthouse illumination, required large permanent magnets to produce their magnetic field. Subsequently, electromagnets were used for this purpose with their current supplied by a smaller "exciter" generator. After 1867, the dynamo-electric, or self-exciting, generator was developed in which part of the generator's working current was fed into the field windings of the electromagnet. With several other significant improvements, the dynamo-electric generator quickly displaced the older generator types and the word "dynamo" gradually became synonymous with "electric generator" (ibid.).

55   Francis B. Crocker, "The History of Electric Lighting," *The Electrical World*, Vol. 27 (9 May 1896), p. 512; and A.J. Lawson, "Generation, Distribution and Measurement of Electricity for Light and Power," *Transactions, Canadian Society of Civil Engineers* [hereafter cited as "Electricity for Light and Power," *Transactions*], Vol. 4 (1890), p. 180.

The first installation of Brush arc lamps was in the Wannamaker Store, Philadelphia, where five Brush dynamos each supplied a four-arc lamp circuit. The first major street lighting installation was in Cleveland, commencing operation on 29 April 1879 with a 12-arc lamp circuit on 18-foot-high poles (John W. Hammond, *Men and Volts, The Story of General Electric* [New York: J.B. Lippencott Co., 1941], pp. 13 and 12).

56   Crocker, "The History of Electric Lighting," pp. 511-512; and Strandh, *A History of the Machine*, p. 157.

57   Hammond, *Men and Volts...*, pp. 27-30, 13 and 23. Some of the early arc lamps were of 3000 and 4000 candlepower mounted on high buildings or on masts as tall as 160 feet to illuminate surrounding blocks of streets (ibid.). Eventually a 2000-candlepower arc lamp became the industry standard. Dynamos for lighting, both arc and incandescent, were rated by the number of lamps on a circuit (Buckley, *Electric Lighting Plants...*, 1894, pp. 82).

58   Hammond, *Men and Volts...*, pp. 23 and 72. As of the mid-1880s, four of the major arc lighting companies had dynamos capable of operating 25-, 28-, 30-, and 40-arc lamp circuits, respectively (ibid.).

The Edison Company claims were "exaggerated," and yet excited a great deal of interest. As has been pointed out by Hughes, it was really inappropriate to compare the efficiency of contemporary high resistance generators designed for arc lamps in series, with the efficiency of Edison's low internal resistance generator designed for incandescent lamps in parallel (Thomas P. Hughes, *Networks of Power, Electrification in Western Society 1880-1930* [Baltimore: John Hopkins University Press, 1983], pp. 37-38).

All the early dynamos were bi-polar, but Edison's was noted for its peculiar field magnets arrangement (Philip Atkinson, *The Elements of Electric Lighting, including Electric Generation, Measurement, Storage and Distribution* [New York: D. Van Nostrand Company, 1890], p. 61).

Experimental incandescent lamps had been produced as early as the 1840s by a number of inventors, and in 1878 an Englishman, Sir Joseph Swan, had developed a reasonably efficient incandescent lamp. Edison's achievement was the development of an extremely durable carbonized thread filament for incandescent lamps, and the marketing on a parallel circuit of a complete lighting system of high efficiency (Herbert W. Meyer, *A History of Electricity and Magnetism* [Cambridge, Mass., MIT Press, 1971], pp. 161-164; and Hughes, *Networks of Power*, pp. 21-32). Commencing in 1883, the Swan lamp was marketed in the United States by the Brush company (Hammond, *Men and Volts...*, p. 57).

59   Meyer, *A History of Electricity and Magnetism*, p. 167.

60   Ibid., pp. 40-46; Hughes, *Networks of Power*, pp. 42-43; and Meyer, *A History of Electricity and Magnetism*, p. 171. Meyer maintains that the Pearl Street Station supplied an area of only about one-sixth of a square mile (ibid., p. 168).

61   "The Montreal Street Railway Power House," *The Canadian Engineer*, Vol. 1 (January 1894), p. 252. As of February 1892, the largest dynamos manufactured in North America had a 275-horsepower rating (Hammond, *Men and Volts ...*, p. 205).

      In Europe, 300-kW dynamos direct-connected to steam engines, were in operation in the late 1880s (John Langton, "Direct Connected Dynamos with Steam Engines," *The Canadian Engineer* [October 1893], p. 163).

62   Hammond, *Men and Volts, The Story of General Electric*, p. 79.

      Battery operated motors had been experimented with as early as the late 1830s. Antonio Pacinotti in 1864 was apparently the first to assert that an electro-magnetic generator could serve as a motor when driven by another generator. Three years later, Dr. Werner Siemens enunciated the principle of dynamo reversibility with respect to his new dynamo-electric generator, but it was not practically demonstrated until the Vienna Universal Exposition of 1873 (Thomas C. Martin and Joseph Wetzler, *The Electric Motor and Its Application* [New York: W.J. Johnston, Publisher, 2nd. ed., 1888], pp. 9-29).

63   Ibid., pp. 62-94; and Hammond, *Men and Volts ...*, pp. 79-84 and 120-138.

64   Passer, *Electrical Manufacturers ...*, pp. 244 and 248; and Elihu Thomson, "Electrical Advance in the Past Ten Years," *Annual Report of the Board of Regents of the Smithsonian Institution*, [hereafter cited as "Electrical Advance, *Annual Report ... Smithsonian Institution*]), Washington: Government Printing Office, (1897), pp. 126-127.

65   These developments are set forth in Hammond, *Men and Volts ...*, pp. 120-138; Martin and Wetzler, *The Electric Motor and Its Application*, pp. 168-179; Passer, *Electrical Manufacturers ...*, pp. 245-246 and 259-260; and Carl W. Condit, "The Pioneer Stage of Railroad Electrification," *Transactions of the American Philosophical Society*, Vol. 67, Part 7 (1977), p. 8.

      In May 1888, the Sprague Electric Railway and Motor Company opened a new 12-mile line in Richmond, Virginia, with 30 cars operating simultaneously over a hilly terrain with sharp curves and steep grades. It represented by far the most ambitious undertaking to that date, and incorporated the latest developments in electric street railway technology. This success attracted a great deal of attention and launched the era of electric traction.

      The critical carbon brushes innovation had been made earlier at the suggestion of Charles Van Depoele, then working for the Thomson-Houston

Company (Thomson, "Electrical Advance," *Annual Report ... Smithsonian Institution*, p. 132).

All of the standard electric street railway technology was in place in 1888, with the exception of the single-reduction motor gear introduced ca. 1890-91 (Passer, op. cit., pp. 259-260).

66   W.G. Ross, "Development of Street Railways in Canada," *Canadian Magazine*, Vol. 18 (January 1902), p. 276; and Lawson, "Electricity for Light and Power," *Transactions*, pp. 231-232.

In 1888, a Thomson-Houston publication showed that electric street cars could be maintained at only one-third the cost of a horse and horse-drawn car on a per-mile basis (Condit, "Railroad Electrification," p. 8).

67   Lawson, "Electricity for Light and Power," *Transactions*, p. 229. Both street railways used the Van Depoele system.

68   After Windsor and St. Catharine's, electric streetcar lines were introduced into Victoria, B.C. (1890), Vancouver (ca. 1891), Ottawa (June 1891), Hamilton (June 1892), Toronto (August 1892), Montreal and Winnipeg (September 1892), Saint John, N.B. (April 1893), Peterborough (ca. 1894), and Halifax (February 1896). See ibid., p. 229; Ross, "Development of Street Railways in Canada," *Canadian Magazine*, p. 276; and "Electrical Department," *The Canadian Engineer*, Vol. 1 (January 1894), p. 249.

69   The major problem associated with the winter time operation of electric streetcars in Canadian cities, where the annual snowfall ranged anywhere from two to 10 feet, was in keeping the tracks cleared (Ross, "Development of Street Railways," p. 276).

70   "The Canadian Ship Canal at Sault Ste. Marie," *Engineering News* (28 March 1895), p. 206. The adjustable split gear was invented by Frank Sprague ca. 1885: see Martin & Wetzler, *The Electric Motor and Its Application*, p. 174.

71   Canal Records, File C-4250/S32-1, Vol. 11, K. Boissevain, Consul General of the Netherlands in Canada, Montreal, to the Hon., The Minister of Public Works, 1 March 1898.

Among those seeking information was Henry Goldmark, an engineer on the American Deep Waterway Commission, who later served as an assistant engineer on the Panama Canal project (ibid., Vol. 12, J. Boyd to C. Screiber, 2 May 1899).

72   Martin and Wetzler, *The Electric Motor and Its Application*, pp. 30 and 114-152; and Passer, *Electrical Manufacturers ...*, pp. 239-240 and 249.

73   SSM Canal Office, Photo Collection, Caption on rock excavation photo of canal construction, n.d.

74   Martin and Wetzler, *The Electric Motor and Its Application*, p. 125; and Edward F. Bush, "A History of Hydro-Electric Development in Canada," (Ot-

tawa: Environment Canada, Canadian Parks Service, manuscript in progress, 1986), p. 52.

75 Strandh, *A History of the Machine*, p. 157; Hammond, *Men and Volts* ..., pp. 209-211; and Daniel Nelson, *Managers and Workers, Origins of the New Factory System in the United States, 1880-1920* (Madison: University of Wisconsin Press, 1975), p. 22.

76 Archibald F. Johnston, *Canadian General Electric's First Hundred Years, A Chronological Sketch* (Toronto: Canadian General Electric Co., 1982), pp. 2-7.

77 Hammond, *Men and Volts* ..., pp. 82-84, 129, and 135-137. Edison had experimented with two electric railway cars on a three-mile track at Menlo Park in 1881-82, and formed the Electric Railway Company of the United States in 1883. But no further effort was put into motor development.

78 Passer, *Electrical Manufacturers* ..., pp. 238-248, and 26-29; and Hughes, *Networks of Power*, pp. 126 and 163.

In 1893, the new General Electric Company also began manufacturing electric locomotives for standard railway service. The first commercial electric railway, as distinct from electric street railway, was the Baltimore, Hampden and Maryland branch of the Union Passenger Railroad Company. It was electrified in 1885 by the Daft Electric Company of Greenville, New Jersey (Condit, "Railroad Electrification," *Transactions of the American Philosophical Society*, pp. 6-8; and Martin and Wetzler, *The Electric Motor and its Application*, pp. 74-76).

Following an 1896 patent exchange agreement with Westinghouse, the General Electric Company would also gain access to the polyphase a.c. motor patented by Westinghouse.

79 Passer, *Electrical Manufacturers* ..., pp. 53-56 and 163.

The Wood arc lighting system patents of the Fuller-Wood company were purchased in 1888. Thomson-Houston acquired the Fort Wayne (Indiana) Company, makers of the "Jenny" arc lighting system, in April 1889, and controlling interest in the Brush Electric Company of Cleveland in October 1889. As of 1890, the Wood apparatus was being manufactured in the Fort Wayne plant, the "Jenny" system was discontinued, and the Brush system manufactured as previously.

80 DRC, *Annual Report*, 1897, p. 120. Subsequently, in 1900, the draw-rods were replaced with new rods of a heavier design (DRC, *Annual Report*, 1900, p. 174).

81 Bush, "A History of Hydro-Electric Development in Canada," p. 6.

82 "The Electric Light, Mr. E.B. Eddy's New Enterprise," *The Canada Lumberman* (1 July 1881), p. 5. A 28-inch hydraulic turbine was belt-connected

to a No. 8, Brush dynamo to power the forty 2000-candlepower arc lamps illuminating a 480 000 square foot area (ibid.). I am indebted to David Lee, Canadian Parks Service historian, for this reference.

83 Montreal was followed by Toronto (a 50-arc light system, fall of 1883), Winnipeg (ca. 1883), and Ottawa (May 1885). As of 1890, Quebec, Halifax, Hamilton, London, Victoria, Vancouver, Saint John, N.B., St. John's, Nfld., Moncton and Sherbrooke in addition to many smaller towns, had replaced gas or coal oil (kerosene) with arc street lighting (Lawson, "Electricity for Light and Power," *Transactions*, p. 182). Ottawa was reputedly the first city in the world to have all its streets lighted electrically (Arthur Porter, "Electric Power," *The Canadian Encyclopedia*, Vol. 1 (1985), p. 557).

84 Bush, "A History of Hydro-Electric Development in Canada," p. 51.

Earlier incandescent lighting installations in Canada were on a very small scale and only temporary. In Montreal, the St. Lawrence Hall and the Bank of Montreal were illuminated by the Maxim system in 1882. Both proved unsatisfactory and were discontinued. In June 1882, the Edison system was introduced in the "Mail" building in Toronto, but this was simply an exhibition of several months duration (Lawson, "Electricity for Light and Power," *Transactions*, p. 183).

The so-called "Maxim system" was not really such. Hiram Maxim, the inventor of the machine gun, had developed an incandescent lamp for the U.S. Electric Company of New York, and it was employed with generators developed by others. Prior to 1885, the U.S. Electric Company was Edison's major rival in the incandescent lighting field, but with only one-fifth as many users as Edison (Hughes, *Networks of Power* ..., p. 100).

From 1881 onwards the Edison Company had made a concerted effort to sell its incandescent lighting system to American textile manufacturers. A decade later, almost every American textile mill of any size had an incandescent lighting system (Hammond, *Men and Volts* ..., p. 63).

85 Lawson, "Electricity for Light and Power," *Transactions*, p. 183-186; and *Heritage Cornwall: Examples of the City's Historic Architecture* (Cornwall: Cornwall LACAC Publication, Vol. 1, n.d. [1979]).

The Canada Cotton Company lighting system was increased to over 1200 lamps in late 1883, and as of 1890 was the largest private installation in Canada. At that date, there were over 12 companies marketing incandescent lighting systems in Canada. The Edison company, with a total of 23 500 lamps installed, and the Thomson-Houston company, with 14 600 lamps, were by far the leading manufacturers (Lawson, op. cit., pp. 183 and 186).

Incandescent lighting not only enabled colours to be discerned properly in textile mills, but was the only form of artificial lighting available during the 1880s that did not present a serious fire hazard (*Electrical Manufacturers* ..., p. 112).

86   Ibid., pp. 184-185; and DRC, *Annual Report*, 1888, p. 101. Later in 1893, the enclosed arc lamp was developed and widely used in stores and large halls (Meyer, *A History of Electricity*, p. 160).

87   Hammond, *Men and Volts* ..., pp. 57-58; and Lawson, "Electricity for Light and Power," *Transactions*, p. 184.

88   Atkinson, *The Elements of Electric Lighting*, 1890, p. 230.

89   DRC, *Annual Report*, 1891, p. 100, and 1897, p. 127. The incandescent lighting system from the Lachine Canal was removed to the Beauharnois Canal, soon to be superseded by the new Soulanges Canal. Ironically, the incandescent lights were used outdoors for illuminating over a mile of the Beauharnois Canal, as well as employed indoors for lighting the workshops and superintendent's office (ibid., p. 101).

90   Ibid., p. 104. In 1892, plans were also made to install arc lamps at the basin and entrance locks of the Rideau Canal in Ottawa (NA, RG43, B61, Vol. 2009, T-2471, p. 372, F.A. Wise to T. Trudeau, 10 March 1892; and ibid., p. 454, Wise to Trudeau, 26 August 1892). Apparently, however, it was only in August 1898 that a seven-arc lamp circuit was installed with incandescent lights in the lock house (ibid., Vol. 2013, p. 344, A.T. Phillips to R. Anderson, 29 December 1898; and ibid., p. 159, Phillips to C. Schrieber, 11 June 1898).

91   One of the first, if not the first, canal to be illuminated with electricity was the St. Mary's Falls Canal at Sault Ste. Marie, Michigan. On 28 June 1884, a Brush arc lamp dynamo driven by a turbine, commenced running a 10-lamp circuit around the Weitzel lock. It replaced kerosene lamps, and the circuit included a lamp in the lock office and two in the machine shop (*Annual Report of the Chief Engineers, U.S. Army, to the Secretary of War for the Year 1885*, Part III [Washington: Government Printing Office, 1885], p. 2111).

92   NA, RG43, B2e, Vol. 1772, Folio 471, William Kennedy & Sons to Spence, DRC, 3 June 1895.

93   A.C. M'Callum, "Turbine Water Wheels," *The Canadian Engineer*, Vol. 1, No. 6 (1893), p. 147; and Hunter, *A History of Industrial Power*, Vol. I, pp. 387-388.

94   NA, RG43, B2e, Vol. 1772, Folio 471, William Kennedy & Sons to Spence, DRC, 3 June 1895.

95   DRC, *Annual Report*, 1897, p. 119.

Leaving aside whatever resistance there was in the canal lighting circuits, both the incandescent and arc lighting dynamos were theoretically large enough. In the Edison 110-volt incandescent lighting system with the standard lamp of 100-Ohm resistance, each lamp generally drew 1.1 amperes. Thus 22 lamps would require 24.2 amperes of current. But a 3-kW, 110-volt Edison dynamo would produce a current of 27.3 amperes. In series arc lighting the system was additive, each lamp requiring 50 volts. Hence, 33 lamps would have required 1650 volts; yet the 40-arc lamp dynamo produced 2000 volts.

96   Joseph P. Frizzel, *Water Power, An Outline of the Development and Application of the Energy of Flowing Water*, op. cit., p. 437. Other sources give slightly different figures for the voltages and power loss in these experiments.

97   As of the mid-1890s, there were a few direct-current high voltage transmission systems in place in the United States and Europe. Voltages as high as 6600 to 8000 to 15 000 volts were transmitted distances of 12 to 20 miles from hydro-electric plants. In the system employed, the high voltage was produced for transmission by a number of dynamos in series and, at the receiving end, motors were driven to run secondary dynamos to produce low voltage electricity for distribution. The system was reasonably efficient, but highly dangerous with a great potential for trouble at the high voltage generating and receiving end of the line (Louis Duncan, "Present Status of the Transmission and Distribution of Electrical Energy," *Annual Report of the Board of Regents of the Smithsonian Institution* [hereafter cited as "Present Status," *Annual Report ... Smithsonian Institution*] Washington: Government Printing Office, 1896, pp. 215-216 and 219).

98   In Canada by far the longest direct-current transmission system was an eight-mile line from the Montmorency Falls to Quebec City. It opened in October 1887 to supply an arc street lighting system, but heavy power losses were experienced. In 1889, generators were installed for transmitting single-phase, alternating current for domestic incandescent lighting in Quebec. All transmission problems, however, were not overcome until 1894-95 when a new powerhouse was built with three 500- kW generators producing two-phase alternating current for transmission at 5500 volts ("Montmorency Falls Electric Plant," *The Canadian Engineer*, Vol. 4 [June 1896], p. 50; and Roos, "Working Paper on Hydro-Electric Technology," p. 211).

99   Slow speed steam engines were used to run the arc lighting dynamos, with high speed engines for incandescent lighting dynamos. Overall as of 1890, a total of 14 750 horsepower was being generated by steam engines in Canada for electric lighting and electric traction applications. Hydraulic turbines generated another 4250 horsepower, or just over one-fifth of the total

electrical power generated in isolated plants and central stations (Lawson, "Electricity for Light and Power," *Transactions*, pp. 181 and 187-188).

100   Martin and Wetzler, *The Electric Motor and Its Application*, pp. 43-44.

101   Atkinson, *The Elements of Electric Lighting*, p. 242; and Duncan, "Present Status," *Annual Report ... Smithsonian Institution*, p. 211.

102   Hughes, *Networks of Power ...*, pp. 83-84. The advantages of the three-wire system, first proposed by a British inventor John Hopkinson, are explained in Atkinson, *The Elements of Electric Lighting*, pp. 242-245.

103   Lawson, "Electricity for Light and Power," *Transactions*, p. 190; and John H. Dales, *Hydroelectricity and Industrial Development, Quebec 1898-1940* (Cambridge, Massachusetts: Harvard University Press, 1957), p. 15. In New York alone, the Edison company had 36 central stations for its incandescent lighting installations (ibid.).

      Hammond (*Men and Volts ...*, pp. 107-108) states that electrical energy, or power, for incandescent lighting could be transmitted up to three miles before it became prohibitively expensive owing to the expense of increased copper in the wiring, and that it could be supplied economically to an area as great as 16 square miles (four miles by four miles). Presumably these figures were based on feeders and/or boosters being used.

104   Duncan, "Present Status," *Annual Report ... Smithsonian Institution*, p. 211. The economical range of direct-current transmission could be doubled and the distribution area quadrupled through employing storage batteries as boosters to improve the load factor. The range could also be increased dramatically by running booster dynamos from motors connected to feeders from the main dynamo, in a system analogous to the high voltage direct-current transmission systems described above (ibid.).

105   C. Coerper, "The Electric Lighting of the North Sea and Baltic Canal," *The Electrical Review*, Vol. 36, No. 917 (21 June 1895), p. 763.

      A Brush arc lighting system, employing a 40-light dynamo, had operated a 30-arc lamp circuit at full brilliance from ten miles away (Hammond, *Men and Volts ...*, p. 107). Hammond, however, does not state the length of the distribution circuit, as distinct from the distance given for the point to point transmission.

106   On the 100-mile-long Suez Canal, electric arc lighting for nighttime operation had been introduced in 1886, almost doubling the carrying capacity of the waterway. Its length ruled out stationary lamps, and consequently a portable lighting plant consisting of a 65-volt dynamo, small steam engine, and two arc lamp search lights, were put on board each ship and temporarily connected to the ship's steam plant during its passage through. The lamps, one forward of 12 000 candlepower and one on the bridge of 6000 candlepower,

were capable of illuminating 1200 metres ahead of the ship and 200 metres all round (R. Percy Sellon, "Electric Light Applied to Night Navigation upon the Suez Canal," *The Telegraphic Journal and Electrical Review* [14 September 1888], pp. 279-282).

107 The North Sea and Baltic Canal, which opened in June 1895 — just over two months before the Canadian 'Soo' Canal — was also completely illuminated by electric lighting. There, however, the new alternating-current technology was employed with dynamos generating 2000 volts which transformers stepped up to 7500 volts for transmission, and stepped down to 25 volts for each lamp. These were spaced 80 to 250 metres apart, on four separate 98-kilometre circuits, flow and return. Both arc and incandescent lamps were run off each circuit. Two central stations were built, containing two steam-driven dynamos of 200 horsepower each. A single dynamo presumably ran each of the lighting circuits. The tidal locks at either end of the canal were hydraulically operated (Coerper, "The Electric Lighting of the North Sea and Baltic Canal," pp. 763-767).

108 In adding a parallel circuit of incandescent lamps to an arc lighting series circuit, it was necessary that they be in balance to preserve the constant-current system. In effect, with the canal arc lamps each drawing 9.5 amperes of current, the incandescent lamp circuit would have had to draw a total current equal to each arc lamp: viz. each of the 22 incandescent lamps would have had to draw a 4/10 ampere current (Atkinson, *The Elements of Electric Lighting*, pp. 241-242; and Hammond, *Men and Volts* ..., p. 58). But the Edison incandescent lamps were designed for a 1.1 ampere current on a 110-volt circuit.

109 NA, RG43, B2e, Vol. 1772, Folio 471, G. Madden, C.G.E., Toronto, to J.B. Spence, DRC, 29 February 1896, 11 March 1896, 1 May 1896, and 9 July 1897. The C.G.E. 800 was a single-reduction-gear motor developed in the United States and first marketed in February 1893 for streetcar use. It was a light, totally enclosed, and waterproof motor (Passer, *The Electrical Manufacturers* ..., pp. 260-261). Direct current motors could be operated off an alternating-current system if a rotary convertor was used.

110 NA, RG43, B2e, Vol. 1772, Folio 471, William Kennedy and Sons to Spence, 2 November 1897 and 8 November 1897; and ibid., J. Boyd, Superintendent, Sault Ste. Marie Canal, to Spence, 18 January 1898.

As of 1897, the older North American practice of belt-driven dynamos had given place to a direct-connected system mounting dynamos directly on the shaft of the hydraulic turbine, or steam engine, providing the power (Thomson, "Electrical Advance," *Annual Report ... Smithsonian Institution*, p. 132).

111   NA, RG43, B2e, Vol. 1772, Folio 471, J. Boyd to Spence, 8 February 1898 and 8 March 1898; and DRC, *Annual Report*, 1898, p. 121.

112   NA, RG43, B2e, Vol. 1772, Folio 471, Canadian General Electric to J. Boyd, 23 December 1897; and ibid., Boyd to Spence, 8 February 1898, and ibid., Boyd to Spence, 16 March 1898.

Arc lighting dynamos were almost invariably series wound with the entire current being passed through the armature coils, field-magnet coils and external circuit in a single path. Incandescent lighting dynamos were either compound wound or shunt wound.

A shunt wound dynamo had two distinct circuits: the external lighting circuit of coarse, low-resistance wire; and a shunt circuit of fine, high-resistance wire wound on the field magnets. The current divided at the upper brush with a small percentage passing through the shunt circuit to excite the field magnets. The compound wound dynamo combined the series and shunt system. It had two distinct circuits but the entire current, rather than a small percentage, flowed through the field magnet coils.

The shunt dynamo had one major advantage. Any increase in the resistance, and consequent drop in current, on the external lighting circuit caused the shunt circuit to increase its current strength, thereby increasing the current also in the external circuit to partially overcome the increased resistance (Atkinson, *The Elements of Electric Lighting*, pp. 34-37).

113   "The Multiphase System of Electricity," *The Canadian Engineer*, Vol. 1 (July 1893), p. 65. Direct-current generators had also been required in industry for electro-plating and for producing copper printing surfaces for printing rollers. Among arc lamps operating on a.c. were the Jablochkoff Electric Candle, the Jamin Electric Candle, and the Sun Lamp (Atkinson, *The Elements of Electric Lighting*, pp. 172-176).

114   Passer, *The Electrical Manufacturers ...*, pp. 131-134; and Thomson, "Electrical Advance," *Annual Report ... Smithsonian Institution*, p. 128. In 1884, the same system was used successfully on a 50-mile series circuit in Italy (Meyer, *A History of Electricity and Magnetism*, pp. 181-182).

Lucien Gaulard and his business partner, John D. Gibbs, were the first to use the induction coil or transformer specifically to obviate the transmission limitations inherent in d.c. systems. In their system, the transformers were connected in series on a high-voltage a.c. primary circuit, and used to lower the voltage to supply the different loads required for incandescent and arc lamps on secondary parallel circuits. (On the development of the transformer in Europe, and subsequently in the United States, see Hughes, *Networks of Power ...*, pp. 86-123).

115 George Westinghouse had founded the Westinghouse Air Brake Company in 1869 to produce his railway air brake, and in 1881 had formed the Union Switch and Signal Company to produce railway switches and signals. In 1884 the latter company entered the incandescent lighting field. The Westinghouse Electric Company was set up in January 1886.

116 Passer, *The Electrical Manufacturers* ..., pp. 138-148.

In the Gaulard and Gibbs system, the constant voltage generator on the high-voltage primary series circuit was initially counted on to maintain the constant current required for arc lamps on the secondary parallel circuits (Atkinson, *The Elements of Electric Lighting*, pp. 246-250). This did not work well, and subsequently various regulators were put on the secondary circuits for operating both arc and incandescent lamps, but with indifferent success.

In 1884-85 an Hungarian electrical company, Ganz and Company, improved the Gaulard and Gibbs transformer and connected the primary circuit in parallel with both arc and incandescent lamps on the parallel secondary circuits. With a constant voltage generator and all circuits wired in parallel, the system was self-regulating.

William Stanley, following the lead of Ganz and Company, improved both the transformer and a.c. dynamo or alternator, and in 1886-87 produced a commercially viable a.c. incandescent lighting system for the Westinghouse company (Hughes, *Networks of Power* ..., pp. 95-106).

To overcome the inherent instability of arc lamps when placed in parallel on constant voltage a.c. circuits, a "resistance" (or ballast) was placed in series with each lamp ("Lighting - Arc lamps," *Encyclopaedia Britannica*, 11th Edition, 1910-11, pp. 665-666).

117 Ibid., pp. 164-175; and Hughes, *Networks of Power* ..., pp. 106-107.

The first long-distance — in contemporary terms — a.c. transmission system in the United States was built by Westinghouse at Telluride, Colorado in 1891 to transmit at 3000 volts over four miles. Two years later, another Westinghouse a.c. system was built at Pomona, California, to transmit 10 000 volts some 35 miles (ibid., pp. 279 and 287).

118 Lawson, "Electricity for Light and Power," *Transactions*, p. 184.

119 Leo G. Denis, *Electric Generation and Distribution in Canada* (Ottawa: Commission of Conservation, 1918), p. 167.

120 Lawson, "Electricity for Light and Power," *Transactions*, pp. 181-186.

121 On the development of Nikola Tesla's induction motor and its characteristics, see Passer, *The Electrical Manufacturers* ..., pp. 278-280 and 302-303; Hughes, *Networks of Power* ..., pp. 112-120; and Martin and Wetzler, *The Electric Motor and Its Application*, pp. 264-267. As Tesla pointed out the existing d.c. system was illogical. Generators produced a.c. which a com-

mutator converted to d.c. for distribution, and then another commutator converted it back again for running each motor.

The patents issued to Tesla in May 1888, covered two basic types of a.c. motors: the synchronous; and the induction or asynchronous. The induction motor, however, eventually became the accepted type for most a.c. power installations. The two motor types are differentiated in Hughes, *Networks of Power* ..., p. 111.

122 Passer, *Electrical Manufacturers* ..., pp. 296-305.

The first, highly innovative, polyphase a.c. power and lighting plants were installed by General Electric at mills in Columbia and Pelzer, South Carolina, in 1894 and 1895 respectively (ibid., pp. 303-305). This system was used by General Electric for an installation, completed in March 1897, at the Canada Cotton Company mills at Valleyfield, Quebec, on the St. Lawrence River — the first electrically powered mill complex in Canada (F.C. Armstrong, "Three-Phase Transmission," *The Canadian Engineer* (March 1897), pp. 322-323; and "A Canadian Cotton Mill," *The Canadian Journal of Fabrics* (December 1897), pp. 367-370. I am indebted to Alan McCullough, Canadian Parks Service historian, for the latter reference.

123 Passer, *Electrical Manufacturers* ..., pp. 281-282; and Hughes, *Networks of Power* ..., pp. 122-125.

A huge 300-horsepower induction motor and an a.c./d.c. generator were also part of the electrical power and lighting system on display at Chicago.

124 "Water Power at Niagara Falls," *The Canadian Engineer*, Vol. 2 (November 1894), pp. 198-199; and Dales, *Hydroelectricity* ..., pp. 19-20. The first, experimental, long-distance transmission of polyphase a.c. had taken place between Lauffen and Frankfurt, Germany in 1891 -- a potential of 30 000 volts sent 110 miles with a 77 percent efficiency. Knowledge of this achievement apparently also had some influence on the choice of polyphase a.c. for the Niagara Falls project (Passer, *Electrical Manufacturers* ..., p. 287).

125 Dales, op. cit., pp. 13, 17 and 20; and Thomson, "Electrical Advance," *Annual Report ... Smithsonian Institution*, pp. 127-131. As late as 1892, the largest a.c. generator operating in North America had a rating of only 150 horsepower (Dales, op. cit., p. 20). A 2100-horsepower dynamo, however, had been displayed at the Chicago World's Fair in 1893 ("Water Power at Niagara Falls," op. cit., p. 199).

On the impact of the American Niagara project on turbine layout and design in North America, see Hunter, *A History of Industrial Power, Vol. I*, pp. 389-394; and Roos, "Working Paper on Hydro-Electric Technology, pp. 50-53.

126 Duncan, "Present State," *Annual Report ... Smithsonian Institution*, 1896, p. 216.

127 Thomson, "Electrical Advance," op cit., 1897, p. 127.

128 Bush, "A History of Hydro-Electric Development in Canada," pp. 5-10; and Roos, "Working Paper on Hydro-Electric Technology," pp. 211-212. See also "Montmorency Falls Electric Plant," *The Canadian Engineer*, Vol. 4 (June 1896), pp. 50-53; and Frizell, *Water-Power*, "Lachine Rapids Power-house," pp. 450-456.

The 1200-horsepower plant at St. Narcisse, transmitting 11 000 volts over 17 miles to Trois Rivieres, was reputedly the first major long-distance transmission system in the British Empire (Dales, *Hydroelectricity*, p. 50). The first polyphase a.c. installation in Canada was apparently a 600-horsepower plant built ca. 1894 at St. Hyacinthe, Quebec, transmitting 2500 volts some five miles (Duncan, "Present Status," *Annual Report ... Smithsonian Institution*, p. 219.

129 DRC, *Annual Report*, 1901-1902, pp. 154 and 189; and ibid., 1904-1907, "Welland Canal."

At a very early date, Gaulard and Gibbs had recognized that their a.c./transformer system was ideally suited for illuminating canals. They recommended that the 100-mile-long Suez Canal be illuminated with their system, transmitting 40 000 volts over the primary series circuit (Hughes, *Networks of Power ...*, p. 91).

130 C.R. Coutlee, "The Soulanges Canal Works, Canada," *Engineering News* (11 July 1901), pp. 30-32; L.A. Herdt, "The Use of Electricity on the Lachine Canal," *Transactions of the Canadian Society of Civil Engineers*, 24 March 1904, pp. 161-170; and ibid., F.H. Leonard, "Electrical Equipment for Cornwall Canal," 24 March 1904, pp. 173-187.

On the Soulanges Canal, a draw-push bar rack was used to work the gates, and a vertical rack worked the stoney valve on each wall sluice. On the Lachine and Cornwall canals, electric motors and new gearing were installed to work the existing gate and valve operating mechanisms.

131 John W. Upp, "The Electric Operation of the Panama Canal Locks," *Sibley Journal of Engineering* (March 1914), pp. 211-214; and David G. Mc-Cullough, *The Path between the Seas, The Creation of the Panama Canal, 1870-1914* (New York: Simon and Shuster, 1977), pp. 599-605.

132 As late as 1913, a d.c. system was installed in Boston to operate a set of parallel locks, an adjacent drawbridge, and waste weir as well as provide the canal lighting and heating system. Electricity was obtained from existing d.c. electric railway and lighting systems on either side of the harbour ("Electrical Equipment of Charles River Locks, Extensive electric motor application

at the lower end of Charles River Basin for operating boat locks connecting the basin with Boston Harbor," *Electrical World* (21 June 1913), pp. 1355-1361).

133 "The Canadian Soo and the Great Canal," *Globe*, 26 October 1895.

134 DRC, *Annual Report*, 1896, p. 97.

135 Charles Moore, ed., *The Saint Mary's Falls Canal*, pp. 31-32; and James O. Curwood, *The Great Lakes, The Vessels that Plough Them: Their Owners, Their Ships, Their Sailors, and Their Cargoes* (New York: G.P. Putnam & Sons, 1909), pp. 28-29 and 39.

136 Heisler, *Canals of Canada*, p. 144. After the founding of the Algoma Iron, Nickel and Steel Company of Canada at Sault Ste. Marie in 1901, an increasingly large tonnage of coal was shipped through the Canadian canal to the Algoma works ("Traffic at Sault Canal Shows Heavy Increase over 1912," *The Sault Daily Star*, 13 September 1913). Heavy shipments of Canadian iron ore were also shipped through the Canadian canal to blast furnaces at Midland and Hamilton, Ontario, following the opening in 1901 of the Helen Mine north of Lake Superior (Margaret Van Every, "Francis Hector Clergue and the Rise of Sault Ste. Marie as an Industrial Centre," *Ontario History* [September 1964], pp. 195-198).

137 MacGibbon, *The Canadian Grain Trade*, pp. 29-30, 42-43 and 55. The quotes are from J.C. Sing, "Canada and Her Waterways," *The Canadian Engineer* (22 January 1909), p. 142, and Curwood, *The Great Lakes ...*, 1909, p. 60, respectively.

The population on the Canadian prairies increased from 250 000 in 1891 to over 1.3 million as of 1911 (W.T. Easterbrook and H.G.J. Aitken, *Canadian Economic History*, op. cit., p. 401).

138 Curwood, *The Great Lakes ...*, p. 62. In 1904, for example, 39 229 553 bushels of grain were shipped through both the Canadian and American 'Soo' canals, exclusive of 68 321 288 bushels of wheat and 5 772 719 barrels of flour (Moore, *The Saint Mary's Falls Canal*, p. 202).

139 Osborne and Swainson, *The Sault Ste. Marie Canal*, p. 104.

140 "Great Canals of the World," *The American Architect* (January-March 1902), p. 47.

141 Osborne and Swainson, *The Sault Ste. Marie Canal*, pp. 104 and 106.

142 Ibid., pp. 105-107. During these years, Canadian shipbuilding boomed as well. In 1902, ten Welland Canallers were under construction in Canadian shipyards as well as a number of the large upper lakes ore carriers (DRC, *Annual Report*, 1902, p. 150).

As related previously, on 1 September 1892 a toll of 20 cents a ton had been imposed on Canadian ships passing through the American St. Mary's

Falls Ship Canal. This was in retaliation for a Canadian rebate system which imposed a similar 20-cent toll on all ships passing through the Welland Canal, but provided for a rebate of 18 cents a ton if they proceeded to Canadian ports for unloading into the St. Lawrence transport system (see Essay I, *Constructing the Canal and Lock, 1889-1894*).

The American toll played a critical role in speeding up construction work on the new Canadian canal, but subsequently was not a factor. In February 1893, the Canadian government had passed an order-in-council revoking the allegedly discriminatory toll rebate system on the Welland Canal, and the American government responded directly by revoking their tariff on Canadian ships using the St. Mary's Falls Ship Canal. Thereafter, there were no tolls on the American canal, or on the new Canadian canal following its opening in September 1895.

143 Keefer, "The Canals of Canada," p. 41; J.W. Le B. Ross, "General Design of a Lock and Approaches," *Journal of the Engineering Institute of Canada*, Vol. 3 (1920), p. 386; and "The Canadian Soo and the Great Canal," *Globe*, 26 October 1895, p. 2.

The last of the wooden schooners were built about 1890, and thereafter they gradually faded out of use (Barry, *Ships of the Great Lakes...*, op. cit., pp. 142-147).

144 Barry, *Ships of the Great Lakes ...*, p. 173. In 1905, 33 vessels were launched on the upper lakes, with a further 40 on order for the next year (Moore, *The Saint Mary's Falls Canal*, p. 202).

145 "The Canadian Canal at Sault Ste. Marie," *Engineering News* (20 June 1895), p. 399; DRC, *Annual Report*, 1897, p. 119; and Sing, "Canada and Her Waterways," *Canadian Engineer* (22 January 1909), p. 142.

146 DRC, *Annual Report*, 1910, p. 262; and Hatcher and Walter, *A Pictorial History of the Great Lakes*, op. cit., p. 269. The smaller Weitzel lock had only 17 feet of water on the sills.

147 Ibid., p. 269.

148 Ross, "General Design of a Lock," pp. 384-385; and "Third Lock at Sault," *The Marine Review* (December 1914), p. 473.

149 Ross, "General Design of a Lock," p. 383; "The Accident at the Canadian Lock," *Engineering Record* (19 June 1909), p. 789; and DRC, *Annual Report*, 1910, p. 266.

150 Canal Records, File C-4250/S32-1, Vol. 11, J.R. Miller to C. Schreiber, 28 January 1897; DRC, *Annual Report*, 1901, p. 210 and 1911, p. 24. The solid timber gates specifications are in: NA, RG43, B2e, Vol. 1771, Folio 1771, DRC, Specification, 18 September 1900.

The solid timber gates installed in 1900-01 and 1910, required only minor repairs over the years and were not replaced until 1960 (Personal Communication, Norman Rutane, Chief of Interpretation, Sault Ste. Marie Canal, to R.W. Passfield, 11 September 1986).

Navigation was also improved by extending the upper and lower approach piers over 800 feet, and by replacing the C.P.R. bridge in 1899 with a longer swing span that did not require a pier in mid-channel (DRC, *Annual Report*, 1899, 1904 and 1905, "Sault Ste. Marie Canal").

151   DRC, *Annual Report*, 1908, p. 168. It was found that fully loaded ore carriers sagged as much as six inches at mid-ship, accounting for the problem (ibid., 1910, p. 267).

152   Canada, *Report of the Auditor General for the Year Ended June 30, 1906*, Vol. 3 (Ottawa: King's Printer, 1906), p. W-118, "Sault Ste. Marie Canal: Repairs."

153   DRC, *Annual Report*, 1900, p. 174; and ibid., 1911, p. 24.

154   Canal Records, Sault Ste. Marie Canal, File C-4272-S32, Vol. 5, J.H. Cravers, Mechanical Superintendent, St. Catharine's, to A.M. Luce, 15 June 1964.

155   Ibid., Vol. 4, W.A. O'Neil to J.D. Bouchard, 2 October 1963.

156   Canada, *Report of the Auditor General ... 1921*, Vol. 3, p. W-108, "Sault Ste. Marie Canal: Repairs"; and Sault Ste. Marie Canal Office, *Ross Notebook*, p. 136, "Power House Second Floor."

157   DRC, *Annual Report*, 1906, p. 183.

158   "Third Lock at Sault," *The Marine Review* (December 1914), pp. 468-473; and "New Canal and Locks at 'The Soo'," *Engineering News* (5 March 1914), pp. 512-519. Each gate was electrically powered by means of gate cables worked by a winding drum driven by a 15-horsepower induction motor, worm gear reducer and spur gears. The floor sluice valves on the new Davis and Sabin locks were hydraulically operated.

On the Panama Canal, see David G. McCullough, *The Path Between the Seas*... The Panama Canal locks, however, had 49 feet of water on the sills.

159   Osborne and Swainson, *The Sault Ste. Marie Canal*, pp. 104-107. In 1925 the St. Lawrence canals and Welland Canal passed 6.21 and 5.64 million tons of freight, respectively.

160   MacGibbon, *The Canadian Grain Trade*, p. 55.

161   Vernon C. Fowke, *The National Policy and the Wheat Economy*, p. 257.

162   Osborne and Swainson, *The Sault Ste. Marie Canal*, p. 107.

163   Albert G. Ballert, "Commerce of the Sault Canals, *Economic Geography*, Vol. 33 (April 1957), p. 135.

164   DOT, *Annual Report*, 1938-39 through 1955-56.

165 "The MacArthur Lock at Sault Ste. Marie," *Engineering* (28 August 1944), pp. 321-323.

166 Ballert, "Commerce of the Sault Canals," pp. 135-137. In 1954, the Suez Canal overtook both 'Soo' canals in annual shipping tonnage. In 1955, for example, tonnage was 118 million tons on the Suez Canal, 114 million at the 'Soo', and 50 million on the Panama Canal (Albert G. Ballert, "The Soo versus the Suez," *Canadian Geographical Journal* (November 1956), pp. 160-166).

167 Hatcher, *Pictorial History of The Great Lakes*, p. 272; and Canal Records, File C-4272/S32, clipping, "New Lakes Era Marked by Freighter," *Globe and Mail*, 4 May 1972. During the late 1950s, the river channels were deepened to provide a 27-foot draught at the 'Soo' to match the new St. Lawrence Seaway system.

168 Canadian Parks Service, Agreements in Recreation and Conservation Branch [cited as ARC Branch], "The Canadian Sault Lock — A Preliminary Discussion of Proposed Options," 21 September 1977, manuscript on file, pp. 1 and 4.

169 Discussion Paper, Minister of Transport/Minister of Indian Affairs and Northern Development, "Future of the Canadian Lock and Adjacent Federal Land at Sault Ste. Marie," 20 January 1978.

170 Various options were discussed previously: see ARC Branch, "The Canadian Sault Lock - A Preliminary Discussion of Proposed Options."

171 The only problems were in August 1930 when the clutch hub on Generator No. 1 broke, necessitating Generator No. 2 being used for the rest of the navigation season, and in 1931 when the large screw of a gate machine broke (DRC, *Annual Report*, 1931, p. 70 and 1932, p. 83).

172 Ibid., 1942-43, p. 19 and 1943-44, p. 20.

173 Canal Records, File C-4272/S32, Vol. 7, F. Roseman, Electrical Engineer, St. Lawrence Seaway Authority, Memorandum, 27 August 1968.

174 Ibid., Vol. 4, W.A. O'Neil to J.D. Bouchard, 2 October 1963.

175 Ibid., Vol. 5, J.H. Travers, Mechanical Superintendent, St. Catharines, Ontario, Memorandum, 15 June 1964.

A No. 3 Riva Governor was installed on the 210-horsepower turbine (Sault Ste. Marie Canal, Canal Office, Case A, Drawing 50053A, "General Drawing of the 39-36 Wheel," n.d.; and ibid., Drawing 50053B, "General Drawing, Power House Turbine," n.d.).

The 13-inch auxiliary turbine of the incandescent lighting system had been removed during the 1950s (Personal Communication, Norman Rutane, Chief of Interpretation, Sault Ste. Marie Canal, to R.W. Passfield, 2 September 1986).

176   Canal Records, File C-4272/S32, Vol. 4, O'Neil to Bouchard, 2 October 1963.

177   R.W. Passfield, Memorandum, "Sault Ste. Marie Canal, Powerhouse Artifacts," to Terry Smythe, Historical Research Division, Environment Canada, Canadian Parks Service, 4 June 1986.

178   Canal Records, File C-4272/S32, Vol. 7, M.S. Campbell to A.M. Luce, 14 September 1967.

179   ARC Branch, Thomas Kearney to E.O. Stenerson, Canals Engineering Division, "Report on Sault Ste. Marie Lock," 6 September 1977.

180   Canal Records, File C4272/S32, Vol. 8, G.A. Moore, St. Lawrence Seaway Authority, to Basic Structural Steel Fabricators, Welland, Ontario, 15 February 1971; and ibid., Vol. 9, J.D. Bouchard, "Sault Ste. Marie Canal Lock," 4 July 1978.

181   DRC, *Annual Report*, 1933-34, p. 74, and 1934-35, p. 82 and 1935-36, p. 81.

182   Canal Records, File C-4272/S32, Vol. 1, D. Oliver, St. Lawrence Seaway Authority, Requisition, 16 October 1959; and ibid., Glorya Nanne, "Canals: Concrete and Steel Replace Limestone in Soo Lock Walls," *Public Works in Canada* (22 May 1962), pp. 28-29. A cut-off wall was also apparently built at this time. In 1976, one was uncovered extending back 50 feet perpendicular to the north wall (ibid., Vol. 9, J. Korsmit, St. Lawrence Seaway Authority, Memorandum, 4 October 1976).

183   *Annual Report*, 1945-46, (Ottawa: King's Printer), p. 21.

184   Canal Records, File C-4272/S32, Vol. 2, A.W. Bridgewater, Memorandum, "Canadian Lock at Sault Ste. Marie, Ontario," 1 April 1963; ibid., General Plan & Profile of Lock Lower Apron, 18 June 1963; and ibid., General Plan of Sill Layout, 21 June 1963.

185   Ibid., Vol. 5, J.D. Bouchard to S. Hairsine, "Ryertex Bearings: Lock Machinery Sault Ste. Marie Canal," 19 June 1964. The previous bearings — Textolite no. 113 — had operated trouble free since their installation in 1943.

186   Ibid., Vol. 1, J.D. Bouchard, Memorandum, "Failure of portion of Lock Floor," 2 October 1961; ibid., "Repairs to Lock Floor and Anchorages," 15 March 1962; and ibid., "Methods of providing tie-downs for lock culverts and floors," 13 October 1961.

      The iron grating over the filling well entrance to the floor culverts was redesigned slightly and renewed in steel during the winter of 1966-67 (Ibid., Vol. 6, J.D. Bouchard to A.M. Luce, 30 September 1966).

187   Ibid., Vol. 8, "Detail & Locations of Centre Wall Culvert Floor Anchorage," 19 February 1971.

      During the winter of 1971-72, the old anchor bolts were tested and found sound, but many threads on the nuts were almost gone. They were partially welded (ibid., Bouchard to M.H. Rehmn, St. Lawrence Seaway Authority, 20 October 1972). In 1974-75, the anchor bolts were re-stressed and re-welded to the steel straps holding down the floor planking (ibid., J.N. Martin, St. Lawrence Seaway Authority, Authorization of Expenditures, 15 May 1974).

188 "The Canadian Ship Canal Lock at Sault Ste. Marie, Ont.," *Engineering News* (28 March 1895), p. 206.

189 Canal Records, File C-4272/S32, General, J.H. Travers, Mechanical Superintendent, DOT, Memorandum, "Unwatering Pumps," 30 September 1963. In 1963, the old pumps were rated at 6000 c.f.m. at a seven-foot head (ibid.).

190 Ibid., Vol. 2, Wright & Barker, Consulting Engineers, to C. Robitaille, Chief, Engineering Services, Ontario Region, Canadian Parks Service, "Installation of Supplementary Lock Unwatering Pumps," 9 January 1984; and ibid., Vol. 10, "Terms of Reference for Engineering Consultant, Re: Installation of Pumps," n.d. (Fall, 1983). The three vertical turbine pumps were manufactured in 1980 and had been used by the Great Lakes Power Company (ibid.).

191 Problems were experienced with both the north and south pump crown gears during the 1960s, and with the re-worked castings (see ibid., Vols. 5 and 7, correspondence June 1964 through March 1968). Finally, in 1972, new right-angled bevel gears were installed on the drive shafts (ibid., Vol. 8, J. Ter Horst to David Brown Gear Industries (Canada) Ltd., 17 January 1972).

III. THE EMERGENCY SWING BRIDGE DAM, 1895-1985

Introduction

The emergency swing bridge dam at Sault Ste. Marie, Ontario, was erected across the upper channel of the newly completed Sault Ste. Marie Canal in 1895-96. This movable dam was a novel structure, only one of which had ever been constructed previously. Owing to several innovative features, it soon became established as the design prototype and standard for its type. The only swing bridge dam to be operated under extreme emergency conditions, it also proved the practicability of an untested and highly suspect working principle. Few emergency swing bridge dams were ever constructed on ship canal systems, and only one remains extant — the Sault Ste. Marie swing bridge dam.

Swing Bridge Dams: Origin and Concept

In planning for the construction and subsequent operation of canals, engineers had always been concerned about the potential danger of boats or barges striking lock gates and severely damaging or carrying them away with a resultant rush of water through the lock. To negate this danger, where a lock opened onto a lake, a navigable river, or long stretch of canal, an extra set of gates — so-called "guard gates" — were generally constructed just upstream of the upper lock gates.[1] In the event the main gates were destroyed, the guard gates could be closed shutting off the flow of water until the main gates could be repaired or replaced.

During the course of the 19th century, as canals came to be constructed on larger scales with wider and deeper locks and increasingly high lifts, engineers had become seriously concerned about closing guard gates in emergencies where the torrent of water pouring through a damaged lock might well be of such a force as to tear the guard gates away. The first of the large locks, tidal locks for ships entering enclosed harbours or river estuaries, had posed no problem. Guard gates could be closed during periods of still water, either at high or low tide, and any damaged lock gate repaired or replaced.

But where large ship canal locks stepped up into vast bodies of inland water, the situation was entirely different.[2] In such circumstances, the prospect of stemming a torrent of water pouring through a damaged lock had perplexed canal engineers for decades until a potential solution was devised by American engineers during the construction of the world's largest lock, the Weitzel lock, on the St. Mary's Falls Canal at Sault Ste. Marie, Michigan, in 1876-81.[3]

At Sault Ste. Marie, a one-mile-long ship canal had been constructed as early as 1853-55 by the State of Michigan to by-pass St. Mary's Falls where the waters of Lake Superior fell 18 feet in flowing down the St. Mary's River to Lake Huron. This canal had been 11 feet six inches deep by 108 feet wide and equipped with two masonry flight locks, each 350 feet long by 70 feet wide with a nine-foot lift.[4] Guard gates were erected to enable the main lock gates to be readily repaired or replaced on a canal that opened onto Lake Superior: the largest freshwater lake in the world, comprising a surface area of over 30 000 square miles.[5] When the Weitzel lock was constructed parallel to the first set of locks in 1876-81, guard gates appeared woefully inadequate to close off the canal channel should the main lock gates be impaired. The so-called "Colossus of Locks" was of far greater dimensions than any previous ship canal lock.

The huge new lock had a chamber 515 feet long by 80 feet wide, narrowing to 60 feet at the gates, and it was erected on a canal deepened to 16 feet. Moreover the single lock had a lift of 18 feet, almost double the standard lift of contemporary locks.[6] In effect should the lock gates be destroyed, the engineers were faced with the prospect of having to gain control over a head of water 16 feet deep, by 108 feet wide in cross-section, contracting to 60 feet and descending in a torrent through an 18-foot drop at the lock. Guard gates could not be safely closed across such a current. In contemplating this potential crisis one of the American engineers, Alfred Noble, developed the concept of a movable dam for emergency use.[7]

The movable dam, erected on the St. Mary's Falls Canal in 1881, consisted of simply a standard double arm swing bridge, with a strong truss built into its floor, and 23 wickets — each over four feet wide — hinged and suspended horizontally under the floor of one arm of the bridge superstructure. When not in use, the swing bridge dam rested on a pivot pier on the canal wall a short distance upstream from the parallel locks. In an emergency, the bridge was to be swung across the canal, coming to rest against a stop on the far side, and the suspended end of the wickets swung down against a raised sill built across the floor of the 108 foot wide channel (Fig. 62). The wicket frames, consisting of iron girders framed together in pairs, were lowered one at a time. Each contained a swinging shutter or gate of solid timber that when pulled close one after another by means of chains and winches, gradually contracted and cut off the flow of water (Figs. 63 and 64).[8]

The swing bridge dam in no way protected the lock gates from damage, but should it be serious, the emergency dam provided a seemingly practicable means of gaining control over a cascading torrent of water. The basic design feature, and principle of operation, consisted of the dam's capability of gradually reducing a large cross-section of fast flowing water and ultimately stemming the current so that the guard gates could be safely closed.[9] With independently operated wickets, the dam structure would be spared the impact of a cascading wall of water.

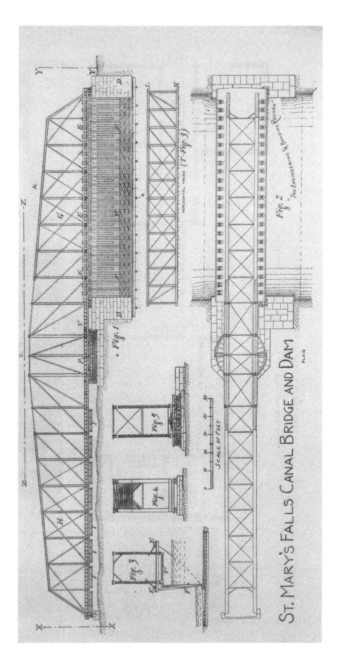

FIGURE 62. Emergency swing bridge wicket dam erected on the St. Mary's Falls Ship Canal in 1881.
(*Engineering and Building Record*, 22 March 1890)

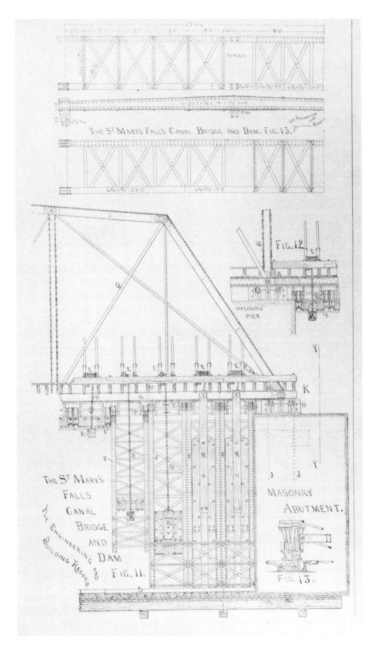

FIGURE 63. Wicket details of emergency swing bridge dam, St. Mary's Falls Ship Canal. (*Engineering and Building Record*, 22 March 1890)

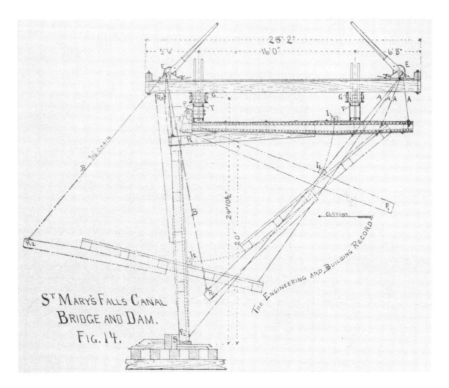

St Mary's Falls Canal
Bridge and Dam.
Fig. 14.

FIGURE 64. Wicket operating sequence, showing the lowering of a wicket frame and pivoting of shutter closed, St. Mary's Falls Ship Canal. (*Engineering and Building Record*, 22 March 1890)

During construction of the Canadian canal at Sault Ste. Marie in 1889-95, the canal engineers of the federal Department of Railways and Canals were acutely interested in the emergency swing bridge dam concept. The Canadian canal, consisting of a one-mile-long excavation and a single lock of 18-foot lift, presented a potential water control problem of even greater magnitude than the American engineers had faced. The 900 foot by 60 foot Canadian lock, with 20 feet three inches of water on its sills, superseded the American lock colossus as the largest lock in the world, and the new canal prism, holding a cross-section of water 152 feet wide by 22 feet deep, was 44 feet wider and six feet deeper than the American canal.[10] Moreover among canal engineers, doubts still remained as to the practicability of emergency swing bridge dams.

As of 1890, the American dam had been in place for five years and was the only structure of its type in service anywhere. It had been operated several times, but only in still water. In each case, the dam was swung across the channel, the wickets lowered, the shutters closed in sequence, and then the water level lowered behind the dam by opening the lock culvert sluices. The dam had proved capable of withstanding the pressure of a 16-foot head of water, but it had never been operated in a strong current. Indeed, it was impossible to test the dam under extreme emergency conditions. Opening the lock gates to let the water rush through was too dangerous an expedient to try under any circumstances and the gates, in any case, could not be opened against a full head of water.[11] Therefore, the emergency swing bridge dam had continued to be regarded as "a practically untried experiment,"[12] and "a curious structure,"[13] the whole concept and application of which reportedly evoked "a good deal of ridicule" among engineers.[14]

Despite the controversy generated by the American swing bridge dam, the Department of Railways and Canals in February 1895 prepared specifications for the erection of a similar structure — but one constructed entirely of steel — on the new Canadian canal which was approaching completion.[15] In March, the contract was let to the Dominion Bridge Company of Montreal on a bid of $75 700, but it was for a structure that incorporated a number of significant improvements in the design and operation of the wickets.[16] Design changes proposed by the chief engineer of the Dominion Bridge Company, George H. Duggan.[17]

The Canadian Swing Bridge Dam: 1895-96

The emergency swing bridge dam was fabricated by the Dominion Bridge Company and erected at the Sault Ste. Marie Canal during August-December 1895.[18] It was supported by a pivot pier, 36 feet in diameter, constructed of concrete and stone masonry alongside the north wall of the canal about 1000 feet above the lock. The bridge superstructure consisted of a double arm, balanced cantilever type of swing bridge with the truss of each arm pin-connected to a central tower over the pivot pier. The bridge was 372 feet long overall, with a 42-foot depth of truss at the centre tower of two trusses set 20 feet apart, centre to centre. The floor consisted of a horizontal truss comprising exceptionally heavy beams at the panel points and heavy cross-bracing to support the lateral pressure on the wickets suspended from the lower chord on the downstream side of the bridge structure. The top chords were connected by lattice girder struts and sway bracing. A plank walkway down the centre of the bridge provided access to the wicket dam's operating machinery, with a 14-foot headroom over the walkway (Figs. 65 and 66).

The swing bridge was of a rim-bearing type, with the superstructure resting on a double drum, loaded at 16 points, and turning about an axis on a circle of roller wheels running on a steel tread (Fig. 67). It was designed to be swung manually by four men.[19] When closed, the bridge rested against two timber stops: one on a raised masonry abutment on the far side of the canal channel; and the other on a raised masonry abutment adjacent to the turntable. The timber stops ensured that the pres-

Fig. 65. The Canadian emergency swing bridge dam, Sault Ste. Marie Canal, closed across the canal during the winter of 1933-34. (Canadian Parks Service, Sault Ste. Marie Canal Office)

sure of the 20-foot head of water acting against the dam would be resisted by the fixed abutments, placing no lateral strain on the bridge turntable (Fig. 68).[20] A con- crete counterweight, in the floor system of the shore arm of the bridge, counter- balanced the wickets suspended from the dam-operating arm of the superstructure.[21]

Twenty-three wickets pivoted in cast steel hinges mounted on the downstream lower chord of the channel arm of the swing bridge. Each wicket frame consisted of two solid girders, three feet deep, set almost six feet apart and interbraced with steel angles and rods to form a hollow box with open bracing on either face. These wicket frames were 40 feet long and, on the bridge being closed, could be swung down against a heavy oak sill raised 12 inches above the bottom of the canal.

The shutters or wicket gates were flat sliding plates consisting of a structural steel framework of 7.125-inch I-beams covered on the upstream face with 1/4-inch- thick steel buckle plates riveted to the frame. Each shutter plate was about 26 feet long and just under six feet wide. The shutter moved freely inside the wicket frame between friction roller wheels fixed to the side girders to guide its travel (Figs. 66 and 68).

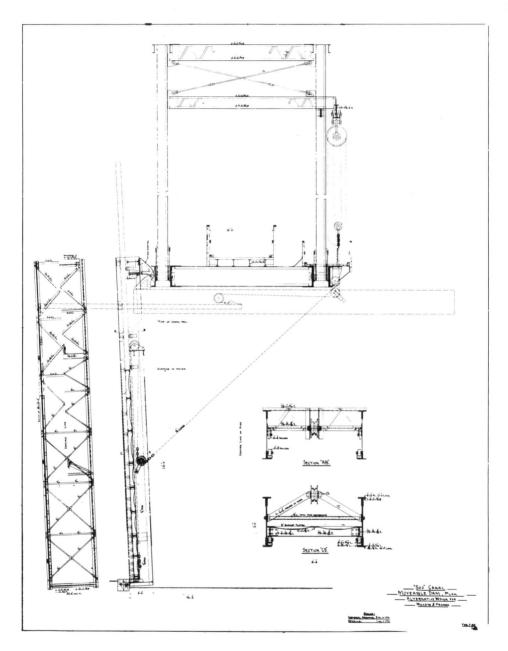

FIGURE 66. The wicket system of the Canadian swing bridge dam, showing manual chain hoist, wicket framing details, and shutter plate in fully closed and fully opened positions. ('Soo' Canal, Movable Dam, Alternative Design for Wickets & Frames, 7 February 1895, Canadian Parks Service)

FIGURE 67. The turntable of the rim-bearing emergency swing bridge dam, showing roller wheels and loading on the drum. (R. Draycott, Sault Ste. Marie Canal Office, January 1983)

FIGURE 68. Timber stops on the swing bridge dam abutments, Sault Ste. Marie Canal. The hinged brackets of the 23 wickets show on the bridge superstructure as well as the shutter plate protruding from the top of each wicket frame. (R. Draycott, Sault Ste. Marie Canal Office, January 1983)

Each wicket frame and sliding shutter was designed to operate as a unit by means of a single chain worked by a manually operated chain-fall hoist. The lower end of the chain was attached to the bottom of the sliding shutter, passed up inside the wicket frame and over a snub pulley mounted near the top of the frame. The chain then passed back down inside the wicket and under another snub pulley mounted at the mid-point of the frame. From there, the chain passed out of the wicket and up to a third pulley mounted on the bottom chord of the upstream truss of the swing bridge. After passing under this third pulley, the top end of the chain was held in a chain stop from which it could be released on being hooked onto the chain fall.

The chain-fall was suspended from an overhead trolley, on a track fixed to the upstream truss of the bridge, and could be moved along from one wicket to the next. When the dam was not in use, the chain of each wicket was simply drawn up by the chain fall and hooked in a chain stop. This held the lower end of the wicket frame suspended horizontally up under the floor of the swing bridge, with the sliding shutter resting on the friction rollers in the upper part of the wicket and protruding out from under the bridge (Figs. 66 and 68).[22]

To operate the wicket dam, once the swing bridge was closed, the chain-fall was hooked to each chain in turn and the chain released from its stop. As the chain was let out by the chain-fall, the wicket frame would gradually be lowered until the force of the water pressed it tightly against the underwater sill. At that point, the sliding shutter plate in the top of the wicket was clear of the water. On the chain being let out further, the shutter would begin to slide down of its own weight inside the wicket frame to close off the flow of water through the wicket. By closing each wicket in turn, the flow of water through the whole 152-foot-wide channel could be gradually closed off.[23]

As late as January 1894, each of the raised abutments on which the timber stops were positioned, was to consist of cut stone masonry. The raised section, approximately six feet high and 13 feet wide by 17 feet long, was to be anchored into its foundation by two large steel chains of flat eyebar links, extending from the rear of the timber stop support to a bar anchorage in the foundation masonry. The foundation was to consist of a massive block of cut stone masonry, 13 feet wide by 38 feet long, carried down over 30 feet from ground level to solid bedrock on a level with the bottom of the canal channel.[24]

Following the extensive use of concrete in constructing sections of the canal prism walls during the summer of 1894, a change was made in the emergency swing bridge dam plans. It was decided to substitute pours of plain concrete for cut stone masonry in constructing the stop support abutments. The flat eyebar chain anchorages, however, were to remain unchanged, and the raised abutments of the stops were to be faced with cut stone. The substitution of concrete would speed construction which was already behind schedule.

With shipping backing up at the American St. Mary's Falls Ship Canal, the new Sault Ste. Marie Canal was opened to commercial navigation on 7 September 1895 with the emergency swing bridge dam not yet operational. This potentially dangerous situation continued through to the close of the navigation on 6 December 1895, and was risked again in the spring following the 21 April 1896 canal opening. Indeed, the concrete pours for the abutments of the swing bridge dam were not completed until mid-May 1896.[25]

The Canadian swing bridge dam was tested thereafter in the only manner feasible. As had been done earlier in testing the American structure, the wickets were closed in the still water of the canal and then the culvert sluices of the lock were opened to run down the water.[26] Closing the dam, however, did not completely seal off the canal channel. Water continued to leak through gaps, almost three inches wide, that had been left between adjacent wickets to prevent their fouling each other on being swung down in a heavy current.[27] But this was of little concern. As stated in the specifications, the primary purpose of the emergency swing dam was simply to reduce the flow of water — in the event the lock gates were carried away — to such an extent as to enable the guard gates to be safely closed.[28]

The Canadian swing bridge dam was the second structure of its type to be erected on a canal, and had the same overall configuration and design function as the earlier American structure. The wicket dam of the Canadian swing bridge, however, was far larger. It closed a canal channel with almost twice the cross-sectional area and, in an emergency, would have to resist correspondingly greater water pressures — a cause no doubt for some apprehension on the part of the Canadian engineers. Besides its unprecedented scale, the Canadian swing bridge dam was also unique in being constructed entirely of steel, and in the novel design and operation of its wickets.[29]

In the American dam, the wicket frames were hinged to the swing bridge superstructure in the same manner as the newer Canadian structure, but the shutter gates worked differently and were of a solid timber construction. The shutters were hinged in the wickets, swinging on trunnions working in journals built into the side girders of each wicket. The journals were positioned about eight feet from the bottom end of the 28-foot-long wicket frames or roughly at the mid-point of the depth of water in the canal with the wicket frames fully lowered.

Operating chains were attached to the upper and lower end of the shutter, and each chain was worked independently by a windlass mounted on either side of the swing bridge superstructure. As the two chains were let out, both the wicket frame and the shutter descended into the water until the bottom of the wicket frame swung against the underwater sill. At that point, the chain attached to the bottom end of the shutter was let out further still, while the chain attached to the upper end of the shutter was drawn in. In this manner, the shutter pivoted in the wicket frame and swung up into a vertical position, closing the wicket (Figs. 63 and 64).[30]

The American wicket dam worked well in still water, but presented a potential problem if operated under emergency conditions. The shutters had to be lowered into the flowing water simultaneously with the wicket frames and then swung closed, under water, with the full force of the current rushing through the wicket frame. The positioning of the shutter pivot at the mid-point of the 16-foot depth of water in the canal rendered the task of closing the wicket somewhat easier. The pressure of the water against the top half of each shutter acting in such a manner as to keep it open, was more than counteracted by the force of water below acting to force the shutter closed. Nonetheless, the perceived difficulty of performing such an operation against a raging torrent of water, may well account for the skepticism expressed by contemporary engineers concerning emergency swing bridge dams.

The Canadian wicket dam avoided potential fouling problems as the shutter gate remained suspended above the torrent of rushing water until the wicket frame seated against the sill on the canal bottom. Moreover, the shutters did not have to be swung under water but simply slid vertically downwards in the wicket frames, guided by the friction rollers. In the Canadian wicket design, instead of pulling against the water current, the operating chain had merely to be let out. The weight of the shutter gate acted to force it down and closed.

Where the working principle of the Canadian wicket dam was concerned some engineers questioned whether the shutters would be sufficiently heavy to overcome the friction in the rollers created by the pressure of the flowing water against the shutters on their commencing to be lowered. But even then, it was realized that the shutters could be forced down with little difficulty using jacks, if need be, working from the swing bridge superstructure.[31]

For more than 14 years, the new Canadian canal operated without difficulty. The swing bridge dam remained on stand by, and was operated occasionally at the close of the navigation season. This consisted of swinging the bridge and lowering several wickets to ensure that the men were familiar with its operation should an emergency arise.[32] And it did. What canal engineers had been dreading for a good many years finally occurred, and in the worst possible circumstances.[33]

The Ultimate Test: June 1909

At noon on 9 June 1909, a Canadian Pacific passenger steamer — the *Assiniboia* — was tied up in the Canadian lock as a large iron ore freighter, the *Crescent City*, entered astern through the upper lock gates preparatory to the two vessels being locked down together. At the same moment a freighter loaded with coal, the *Perry G. Walker*, was proceeding along the lower channel intending to tie up below the lock until the down lockage was completed.[34] Owing to a misunderstanding between the bridge and the engine room, the *Perry G. Walker* failed to reverse engines on approaching the lock. The freighter glanced off the canal wall and struck the south leaf of the lower gates, forcing it back into the lock. The north leaf of the mitre gates, left unsupported, was immediately forced up over the mitre sill and flung open by the pressure of water in the lock chamber.

Within moments, the rush of water out of the lock drove open the south gate leaf, sweeping the *Perry G. Walker* back down the channel. The *Assiniboia* and *Crescent City* shot down out of the lock on a cascading torrent of water, striking and breaking off both leaves of the lower gates in passing.[35] Almost immediately the upper gates were sucked out from their wall recesses, and torn away by the water pouring down into the lock chamber. Out of five sets of gates on the Canadian lock only three remained unscathed. The guard gates above the lock, and another set of guard gates below the lock, had remained securely fastened in their wall recesses. The canal workforce also managed to lash back and save the intermediate gates — a set of auxiliary gates, positioned in the lock chamber just above the lower gates, for use when the lower gates required repairs.[36]

With the upper and lower gates of the lock torn away, catastrophe threatened. The waters of Lake Superior, to a depth of 20 feet, were free to flow down through the canal. A wall of water surged forward in a raging torrent, dropping three to four feet in the upper canal, 12 feet at the breastwall of the upper gates, and a further four feet at the lower end of the lock chamber. The fast flowing water had a surface velocity of about seven feet per second in the upper canal channel, increasing to about 14.6 feet per second in the lock chamber — a calculated flow of 16 956 cubic feet per second (Figs. 69 and 70).[37] Efforts began immediately to activate the emergency swing bridge dam in an attempt to stem the heavy current.

In operating the emergency dam, a number of delays and difficulties were experienced. On swinging the bridge, the hand turning gear broke necessitating the fetching of a team of horses, rope and tackles. As a result, it took almost three hours to close the bridge. Then a number of wicket frames were lowered, after which the chains were let out further allowing the shutters to descend one after another. Most of the shutters descended without any serious difficulty, but operating the chain-fall by hand proved very tedious as more wickets were closed. This work continued

FIGURE 69. Water cascading past destroyed upper lock gates, following the 9 June 1909 accident. The emergency swing bridge dam upstream has yet to be activated. (Young, Lord & Rhoades, National Archives of Canada, PA-146239)

throughout the evening and into the early morning of June 10th. Finally at 3 a.m., all of the wickets were closed with the exception of six with jammed shutters, and another which had its frame twisted by the rushing water. Thereafter, 50-ton hydraulic jacks were used to force down the six shutters (Fig. 71). The twisted wicket frame was swung up out of the way.[38]

To close the six-foot-wide gap left by the removal of the twisted wicket frame, pine timbers — 10 inches by 12 inches by 24 feet long — were placed across the opening, one on top of another, and pushed down into the flowing water with the hydraulic jacks. This proved exceedingly difficult and time-consuming work as the timbers had to be forced down across the protruding rivet heads and brace rods on the upstream face of the wicket frame girders on either side of the gap. When within a foot of the canal bottom, the timber bulkhead resisted all further movement. A timber was then loaded with sand bags and dropped into the current which carried it into place. But with the dam seemingly closed, still another problem emerged. Leakage through the three-inch gaps between the wicket frames far exceeded what had been expected, and the flow of water remained uch too great to risk closing the guard gates (Fig. 72).

FIGURE 70. Water cascading through broken lower gates, following the 9 June 1909 accident. (*Engineering News*, 17 June 1909)

FIGURE 71. Closing the wicket dam, Sault Ste. Marie Canal, 10 June 1909. Water continues to pour through the gaps between wicket frames and through the several wickets with jammed shutters. (Courtesy of AMCA International Ltd. [Dominion Bridge Company Limited], National Archives of Canada, PA-120167)

FIGURE 72. Diminished flow of water through lock at the upper gates, with wicket dam openings yet to be totally sealed off. (Young, Lord & Rhoades, National Archives of Canada, PA-146246)

To reduce the flow, planks were placed over as many gaps as possible and sand bags and straw were dumped against the wickets to seal the smaller gaps around the shutters. This effort reduced the head of water in the lock, but the water still flowed at a rate of 4084 cubic feet per second with a surface velocity of 4.44 feet per second. Nonetheless, after almost 3-1/2 days of unstinting effort to stem the current, on the afternoon of 13 June an attempt was made to close the lock gates.[39]

At this point, it appeared that the intermediate gates down in the lock chamber might be closed with less difficulty than the upper guard gates. Hence heavy timbers were secured to the downstream side of each leaf of the huge, 37 feet wide by 44 feet six inches high, intermediate gates with cables attached to be worked by means of block and tackles anchored to snubbing posts along the lock walls. To counteract the current somewhat, two tugs — the *Rooth* and the *General* — were tied up in the lower end of the lock, and their propellers used to "back water" as the two 87-ton gate leaves were slowly swung out.

Almost immediately, the force of the flowing water acting against the bottom of the framed wooden gates warped and bent them to the point where disaster threatened. But although severely strained, the gates did not split and were successfully closed against the mitre sill.[40] The continuing heavy leakage through the swing bridge dam wickets filled up the canal above the closed gates in only 45 minutes, and the upper guard gates were then closed in still water, bringing the crisis to an end.[41]

Impact and Design Influence

Despite the all but overwhelming difficulties experienced in gaining control over the torrent of water surging through the Sault Ste. Marie Canal lock, the operation was considered an unqualified success by canal engineers cognizant of the magnitude of what had been achieved.[42] Several inadequacies were apparent in the design details and functioning of the emergency swing bridge dam, but modifications could be readily made to rectify these shortcomings.[43] The Canadian experience had proved the feasibility of swing bridge dams under the most trying of emergency conditions. It showed conclusively that such structures could be employed successfully to control a heavy flow of water through a damaged lock.[44]

Such an experience was invaluable at a time when plans were already underway to construct emergency swing bridge dams elsewhere based on the Canadian prototype with its sliding steel shutter wickets. On an enlarged St. Mary's Falls Canal at Sault Ste. Marie, Michigan, and on a new canal under construction at the

Isthmus of Panama, even larger scale potential water control problems could now be approached with confidence.[45]

At Sault Ste. Marie, canal traffic had reached staggering proportions in the two decades following the opening of the large Weitzel lock in 1881, and had continued to increase dramatically.[46] Within a year of the opening of the Canadian canal in 1895, some 16 229 061 tons of freight passed through the Sault Ste. Marie canals, carried in 18 615 vessels during a single seven-month's navigation season. This was twice the tonnage on the Suez Canal, which was open year round, and established both 'Soo' canals as by far the busiest in the world.[47]

With the continuing development of the great iron ore mines of the Mesaba, Goebic and Vermilion ranges of Lake Superior, the phenomenal growth of the grain trade from the American west and the newly opened Canadian west, and the rapid growth of a return trade in coal, ship builders were hard pressed to keep up with the demand for shipping.[48] Ever larger vessels were built to accommodate the bulk cargoes. The largest were constructed for the iron ore and coal industries whose traffic was localized on the upper lakes above Niagara Falls and fed by railway networks on the Lake Superior and Lake Erie ends of the shipping route. Hence where canals were concerned, the largest of the lake vessels used only the facilities at Sault Ste. Marie.[49]

In view of the increasingly heavy freight traffic through the 'Soo', the American government had begun planning soon after the completion of the Weitzel lock for the construction of an even more mammoth structure to replace the two combined locks erected earlier by the State of Michigan. The new Poe lock had been completed in 1896. It was 800 feet long by 100 feet wide with 21 feet of water on the sills, and had superseded the Canadian lock as the largest lock, by volume, in the world.

The American canal prism had also been deepened from 16 feet to 25 feet.[50] The existing emergency swing bridge dam, however had remained in place and operational pending a solution to the problem of how to protect the canal locks. Adjacent to the swing bridge dam the canal was left at its former depth, even though the 16 feet six inches of water on the sill of the emergency dam limited the draught of ships passing through the newly deepened canal.

During the winter of 1898-99, the St. Mary's Falls Ship Canal was deepened to 25 feet throughout with the removal of the sill for the emergency swing bridge dam. This rendered the emergency dam of 1885 inoperable, and it was subsequently removed.

To protect the Weitzel and Poe locks, guard gates were placed in the newly deepened canal channel adjacent to the abutments of the former movable swing bridge dam. The guard gates, for closing off the 108-foot-wide canal in an emergency, worked in the same manner as mitre gates but were built in a semi-circular or arch configuration. Each leaf, of heavily reinforced Oregon fir, had a 60.6-foot chord and a 98-foot radius of curvature. The 31.6-foot high gates were designed to withstand a 27-foot head of water and to overcome the major problem experienced in operating safety gates in an emergency situation: viz. the impact of a wall of water, pouring in a torrent down through a canal, tending to tear away or buckle the gate leafs during attempts to close them. The semi-circular guard gates were far stronger than the conventional guard gate in cross-section, and the curved face of each leaf would tend to deflect the impact of the fast flowing water as the gates were closed.[51]

On the St. Mary's Falls Ship Canal, the revival of the older approach of protecting locks on a canal opening into a large body of water by means of guard gates, albeit of a new design principle, proved only a temporary expedient. Indeed, doubts may well have remained as to the feasibility of operating the semi-circular safety gates in an emergency situation. When work proceeded in 1905-11 to widen the canal channel from 108 feet to 300 feet, the emergency swing bridge dam was reintroduced on the St. Mary's Falls Ship Canal. Plans were prepared for a double-arm swing bridge dam to protect the Weitzel and Poe locks in the event of an accident to the lock gates.

In designing the new emergency swing bridge dam, the American engineers improved on the Canadian wicket dam prototype and overcame several apparent deficiencies in its operation.

St. Mary's Falls Canal Swing Bridge Dam, 1910-11

The American swing bridge dam on which work commenced in 1910 was a standard double arm, riveted truss, structure mounted on the pivot pier in the middle of the enlarged canal. A strong truss was built into its floor, and wickets were hinged to the lower downstream chord in the conventional manner. The wickets, however, were suspended from both arms of the superstructure to enable it to close the 108-foot-wide channels, some 21 feet deep, on either side of the pivot pier island. There were 16 wickets on each arm, each over six feet wide, with a 2-1/4-inch gap between the girder frames. The steel wickets were of the same design as the Canadian structure, but of a somewhat stronger construction with heavier bracing between the side girders and heavier steel — eight-inch I-beams and 3/8-inch plate — in the sliding shutters. The American dam had also been improved in its ease and speed of operation by utilizing roller bearings in the wickets to reduce friction on the shutters being closed, and in applying electrical power to operate the structure (Figs. 73 and 74).

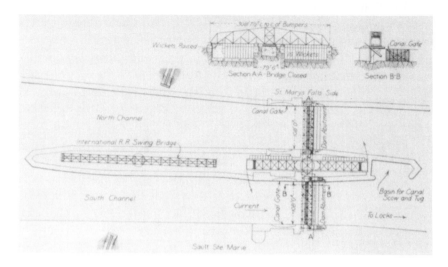

FIGURE 73. Double arm, emergency swing bridge wicket dam erected on an island in the enlarged canal channel, St. Mary's Falls Ship Canal, 1910-11. (*Engineering and Building Record*, 5 August 1911)

Fig. 74. South arm of the double arm emergency swing bridge dam, showing heavily braced wicket frames and sliding shutter plates, St. Mary's Falls Ship Canal. (*Engineering and Building Record*, 5 August 1911).

On the American dam, two 50-horsepower motors were used to quickly swing the bridge in a matter of minutes, and each wicket was raised and lowered electrically. A separate 150-horsepower motor drove a line shaft on each arm of the bridge. The hoists on the individual wickets could be thrown into gear on the shaft to raise and lower the wicket frames and shutters in any order desired. To ensure that the wicket frames seated squarely against the sill on being lowered, wells or pockets were constructed in front of the sill to catch any debris that might wash down the canal.[52] Improvements were also made in the Canadian swing bridge dam, but of a somewhat different nature.

The Improved Canadian Swing Bridge Dam, 1910-11

Following the closing of the guard gates of the Canadian lock on 13 June 1909, the damaged lock chamber was pumped out and temporary repairs made. Spare gates were hung to replace the upper main gates, and using the intermediate gates in place of the missing lower main gates, the canal was reopened for navigation on 21 June, only 13 days after the accident.[53] The swing bridge dam rested on its pivot pier alongside the canal, damaged but still operable if need be. An examination revealed that the one-inch steel rod bracing in the lower panels of the wicket frames was too light. Many rods were bent or broken, and this had allowed four of the wicket frames to buckle somewhat.[54]

With the canal back in operation, the superintendent had been anxious that the emergency dam be fully repaired,[55] but nothing was done until after the close of the 1910 navigation season. Finally tenders were sought, and the Department of Railways and Canals awarded the repair work to the Canada Foundry Company of Ottawa. The wicket dam component of the swing bridge structure was to be rehabilitated at a cost of $7000 and rendered completely operational by 31 March 1911.[56]

In carrying out this work, the badly twisted wicket frame was totally replaced, several frames were straightened, and the rod bracing was cut out of all of the wickets. The rods were replaced by heavier angle and tee bracing set back into the frame clear of the upstream face of the side girders. The strut or diaphragm at the bottom end of the wicket frame was also strengthened.[57] To close off the gaps between the wicket frames, bevelled timbers were bolted to the outside of each wicket frame.[58] The bevelled timbers were arranged so that if the wicket frames were lowered alternately one after the other, the wicket frames in between on being swung down would be driven snugly into place by the force of the water — like a tapered stopper — thereby eliminating the leakage problem experienced in June 1909 (Figs. 75 and 76).

FIGURE 75. Lowering wicket frames for inspection, Sault Ste. Marie Canal, showing heavy angle and tee cross-bracing, substituted in 1911 for the original one-inch-diameter diagonal steel rod bracing in the wicket frames. (R. Draycott, Sault Ste. Marie Canal Office, January 1983)

To overcome the slowness of the manually operated chain-fall in lowering the wickets, a vertical boiler and steam hoist were erected on the swing bridge super-structure. The actual date of the steam hoist installation is uncertain, but appears to have been undertaken as part of the 1911 rehabilitation of the swing bridge dam.[59] The steam hoist operated by means of a 15-inch hoisting drum, and a 3/4-inch-diameter wire rope or cable passing over a series of 10-inch-sheaves to the wicket chains at the chain stops.[60] Why a steam hoist was employed, remains a mystery.

The Canadian lock constructed at Sault Ste. Marie in 1895, had been the first in the world to be operated by electrical power. Both the lock gate and valve machinery were driven by d.c. electric motors utilizing hydro-electric power generated in the canal powerhouse adjacent to the lock chamber, and a.c. power as well was being generated at a nearby private power station.[61] Electrical power was readily available on site, cheap, and instantaneous in its application. In contrast, it

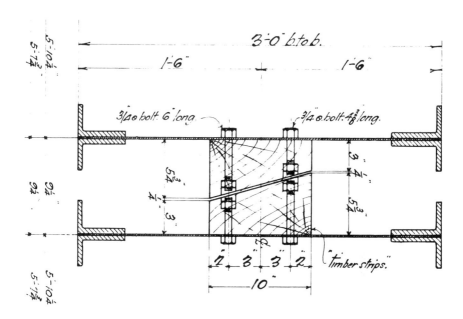

FIGURE 76. Bevelled timber seals, bolted along the outsides of each wicket frame, to close the gap between adjacent wicket frames, December 1911. (Canadian Parks Service, Canal Records, File C-4256/S32-10, Vol. 2)

took over two hours to raise sufficient steam in the boiler to operate the steam hoist.[62] Nonetheless the steam hoist did expedite the operation of the wicket dam, and the improvements made in rehabilitating the swing bridge dam overcame design deficiencies exposed in actual operation under the extreme emergency conditions of June 1909. These deficiencies were carefully studied in turn by the engineers engaged in constructing the Panama Canal.

The Panama Canal Swing Bridge Dams, 1912-14

At the Isthmus of Panama, a decision had been made in 1907 to abandon a sea-level canal in favour of constructing a high-summit level canal with locks.[63] This involved the creation of a huge summit-level lake reservoir, Lake Gatun — over 32 miles long, and covering about 167 square miles — 85 feet above sea level.[64] Three flight locks, the Gatun locks, were required on the Atlantic side to raise or lower vessels and three locks — a single lock at Pedro Miguel and two combined locks at Miraflores — were needed on the Pacific slope. Passage through the canal was to be expedited by constructing all of the locks in pairs, a total of 12 locks, enabling vessels to pass going in either direction.[65] Once again, however, canal engineers had faced the critical problem of controlling a raging torrent of water flowing out of a large lake should the lock gates be destroyed.

At Panama, the concern about vessels carrying away the lock gates was all the more acute as the locks were to be constructed on an enormous scale to take the largest ocean vessels planned for construction — the battleship *Pennsylvania* and an ocean liner, the *Titanic*.[66] Each lock was to be 110 feet wide by 1000 feet long with over 49 feet of water on the sills, over twice the depth of the 'Soo' canals. Should the upper gates be carried away, it was calculated that Lake Gatun would discharge at a rate of 95 000 cubic feet per second and at a velocity of 24.1 feet per second, a torrent of water far surpassing what the Canadian swing bridge dam had faced.[67] The engineers, however, were fortunate in having precedents to draw on in designing large scale lock structures and protecting them from potential damage.

On the Panama Canal, the engineering staff of the U.S. Army Corps of Engineers was thoroughly familiar with the massive locks constructed previously at Sault Ste. Marie, and with the concept and design of the emergency swing bridge dams. Indeed, the American plan for a summit-level canal at the Isthmus was proposed and prepared by Alfred Noble. He drew on his experience at Sault Ste. Marie having served as assistant engineer during the construction of the Weitzel lock, prior to designing the first emergency swing bridge dam ever constructed. Noble was assisted in designing the Panama locks by Colonel Joseph Ripley, a former chief engineer on the St. Mary's Falls Canal. On Ripley resigning his position, the lock design work was taken over by Lt.-Col. Harry F. Hodges, who had worked on the construction of the gigantic Poe lock at Sault Ste. Marie.[68]

These men were well aware of the water control problems faced at the 'Soo', and had studied the design details of the Canadian swing bridge dam focussing on the particular problems experienced in operating the wicket dam during the June 1909 crisis.[69] Indeed, the swing bridge dams designed for the Panama Canal were based on the Canadian emergency dam prototype,[70] but with several significant modifications made to take account of the greater head of water to be controlled at the Isthmus and what had been learned from the Canadian crisis.

On the Panama Canal, which opened in August 1914, several measures were taken to protect the locks. Vessels were not allowed to proceed under their own power near the locks, but were towed by locomotives running along the canal. Heavy fender chains were also strung across the bottom of the canal 100 feet ahead of the upper locks. On being drawn taut in an emergency, the chains were capable of slowing the largest vessel to a halt. These precautions made it highly unlikely any vessel would ever strike the lock gates. Nonetheless in case an accident should occur, heavy steel guard gates were positioned above the head of the single locks at Pedro Miguel and each flight of locks at Gatun and Miraflores, and were to be worked in conjunction with emergency swing bridge dams, six in all.[71]

In designing the Panama Canal structures, a concentrated effort was made to perfect the Canadian swing bridge dam prototype in all its details, overcoming all known shortcomings, and rendering its operation as rapid and smooth as possible.[72]

The superstructure consisted of an unequal arm, centre bearing swing bridge 262 feet long with its main trusses 32 feet apart. As in the Canadian structure, each arm of the bridge consisted of cantilevered trusses pin-connected to a centre tower or post over the pivot pier. The wickets and the horizontal truss on the long arm were counterbalanced by a massive concrete counterweight of 909 tons on the short arm.[73] On the Panama dams, however, the wicket frames were so long — over 59 feet in length — that to counterbalance their great weight the horizontal truss had to be placed on the downstream side of the long arm rather than built into the floor between the main trusses. The great length of the wickets also necessitated the erection of crane-jib girders, extending 30 feet out from the upstream side of the superstructure, on which the machinery for lowering and raising the wicket frames was mounted. The bridge had a manually operated turning system similar to the Canadian structure, but to speed this operation two 110-horsepower a.c. motors were installed. They were capable of swinging the bridge in a matter of minutes (Fig. 77).

The wicket frames were hinged and suspended under the bridge superstructure in the conventional manner on steel hinges and one-inch wire ropes. Each of

FIGURE 77. Unequal arm, emergency swing bridge dam, Panama Canal, 1912-14. Supporting trusses for the long wicket frame girders are cantilevered out on either side of the swing bridge superstructure. (K. Elder, postcard collection)

the six wicket frames consisted of two deep box girders framed together about nine feet apart. The wicket girders were designed to pose as little resistance as possible to flowing water, and swung down into large cast iron pockets built into the face of a reinforced concrete sill. The pockets were 14 feet deep to catch any debris which might otherwise prevent a wicket from seating tightly against the sill, and the sill itself was shaped so that the current of water would wash away any potential obstruction to the shutters being closed. Each wicket frame had its own machinery and a separately controlled motor with variable speed gearing, and could be lowered in just four minutes (Fig. 78).

The shutters consisted of structural steel frames faced with steel buckle or dish plates. Instead of a single full-length shutter, however, there were five plates or roller gates on each wicket. This change was necessitated by the extreme length of the wickets, and the desire to keep the shutters clear of the flowing water on lowering the wicket frames. Rather than travelling in the wicket frame, the plates slid down against the upstream face of the wicket girders. The plates were also wider than the wicket frame, and overlapped to make contact with one another forming a solid wall six panels or plates wide across the canal prism. Roller wheels and flanged wheels were mounted on each panel and ran on the flange of the wicket girders. All of these guide wheels were equipped with roller bearings as the American engineers believed that their absence was a serious defect in the sliding shutters of the Canadian wicket dam (Fig. 79).

To eliminate any potentially serious leakage problem, the upper and lower edges of each plate were wedge-shaped to interlock one with another, and the vertical joints were grooved. Once all of the panel plates were in place, wedges or splines were slid down between them to staunch leakage.[74] All leakage was not prevented by this mode of construction. It would, however, reduce the flow to 950 cubic feet per second, or roughly one percent of the total calculated discharge if the gates were carried away. This was regarded as the practical minimum leakage for a wicket dam of any type, and would render the water below the dam comparatively still.[75]

Each set of five plates on a wicket was operated by a separate electric motor, with another motor operating the wedges. With all of the operating machinery fixed in place at each wicket and electrically powered, the dam could be closed — in still water — in less than an hour.[76] To operate the wicket dam as quickly and as efficiently as possible, a crew of 14 or 15 men was required with two men at each of the six wickets to ensure that the shutter plates engaged properly on the wicket girder rails.[77]

The six emergency dams on the Panama Canal, first tested in May 1914, marked the culmination of the design evolution of the novel emergency swing bridge dam concept introduced a quarter century earlier. Based on what had been

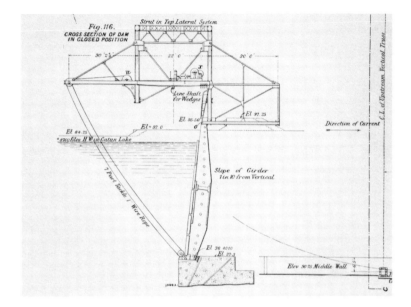

FIGURE 78. Cross-section, Panama Canal swing bridge dam, with wicket frames lowered and sliding shutter plates closed across the face of the wicket frames. (*Engineering*, 18 July 1913)

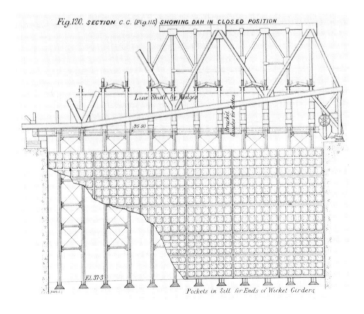

FIGURE 79. Cut-away view of swing bridge dam, Panama Canal, with interlocking shutter panels in place across the face of the box-girder wicket frames. (*Engineering*, 18 July 1913)

learned from the Canadian experience of June 1909, the Panama Canal engineers had no doubt that new structures were capable of operating effectively and efficiently in an emergency to bring any threatening torrent of water under control.[78] But fully developed swing bridge dams were expensive, complicated structures. Each of the emergency dams on the Panama Canal had cost over $430 000.[79]

During the following decade, the emergency swing bridge dams on the Canadian and American canals at Sault Ste. Marie and on the Panama Canal, were regarded as the standard type of movable emergency dam. They were considered to be the only type of structure suitable for erection in the comparatively rare situations where a ship canal stepped up into a large body of water, and consequently required an emergency dam to stem the flow of a raging torrent of water should the lock gates of the upper lock become badly damaged or carried away.[80] But at the same time, canal engineers were concerned about the heavy construction costs and manpower requirements of the emergency swing bridge dams on the Panama Canal. They began seeking other ways to accomplish the same purpose or to negate the need for movable emergency dams.

In less than a decade, three different alternatives to the standard emergency swing bridge dam were developed and employed on newly constructed ship canals. These alternatives involved respectively: the elimination of the emergency swing bridge dam in favour of another safety device; an adaptation of the traditional emergency swing bridge wicket dam; and lastly, the development of a totally different type of movable emergency dam.

Innovation and Adaptation at the 'Soo': 1914-24

At Sault Ste. Marie, the American government had decided as early as 1907 to construct a second canal parallel and adjacent to the existing St. Mary's Falls Ship Canal to meet rapidly increasing shipping demands. When completed the new canal was 4800 feet long with a prism 350 feet wide by 25 feet deep. Two parallel twin locks, each 1350 feet by 80 feet with 24.5 feet of water on the sills, were built: the Davis Lock, opened in 1914; and the Sabin Lock, completed in 1919. Despite the increased depth of navigation over the older canal, no provision was made initially for an emergency swing bridge dam.

To protect the new canal, should the lock gates be struck and carried away, each of the long locks was equipped with two sets of heavy steel-plate girder gates at either end. These dual gates were in addition to the conventional heavy timber guard gates needed for dewatering the locks when repairs were required.[81] Both sets of upper gates, and both sets of lower gates, were designed to work in unison at every lockage. As explained at the time, this was done:

*so that the two levels will always be separated by two closed gates
and the possibility of accident under these conditions is thereby
rendered so remote that it does not seem necessary to further multi-
ply precautions.*[82]

In effect, on a ship or ships entering a lock, two sets of lock gates would be
closed at the far end of the chamber. If one set of gates were struck and driven open
or destroyed, the other set would still be in a closed position preventing the water
from pouring down through the lock. Such was the theory, although untried in practice.

After several years of operation with the novel dual lock gates system, the
situation was re-examined. Freight tonnage on the St. Mary's Falls Canal in an
eight-month navigation season had reached over 60 million tons, one-sixth of the
total annual ton-mileage of all the American railways combined, and traffic was
increasing rapidly. An average of 30 vessels a day passed through the canal at a
time when lake freighters were growing dramatically in size. The largest boats 625
feet by 60 feet approached 10 000 tons and, if out of control, represented a problem
equivalent to stopping four large freight trains. Consequently, the dual lock gates
system no longer appeared to provide protection under all imaginable circumstances — a runaway freighter could conceivably crash through both gates.

In the event a set of dual lock gates were destroyed, with the other set of gates
open, it was deemed critically important to have an emergency device to cut off
the torrent of water flowing down through the lock. Hence the engineers proceeded
to design a movable emergency dam capable of closing off either of the narrow approach channels, each 90 feet wide by 24.5 feet deep, leading into the two parallel
locks.

At first, consideration was given to erecting a double arm, emergency swing
bridge dam, 300 feet long, on the nose pier between the locks. Wickets suspended
under each arm would enable either channel to be closed off in the event of an accident. This proposal, however, was ultimately rejected owing to space limitations
on the nose pier and the realization that if one lock were damaged, the closing of
the swing bridge superstructure would effectively close the canal through block-
ing access to the undamaged lock as well.[83]

To overcome these difficulties, the conventional emergency swing bridge dam
was extensively modified. The result was an adaptation that maintained the design
and operating principle of the movable wicket dam, but eliminated the cumbersome swing bridge superstructure. Two relatively small, and movable, bridge spans
were substituted for the swing bridge: an "operating bridge" and a "wicket bridge,"
with detachable wickets. Both bridge spans, as well as 30 detachable wickets —
the disassembled frames and shutters — were to be stored on the nose pier between
the two parallel locks. To assemble the movable emergency dam across either canal

channel, a stiff-leg derrick was placed adjacent to the head of each lock. Both derricks were steam powered and had a 53-foot-high mast and 80-foot-long boom, capable of lifting and swinging the 82-ton wicket bridge span and 20-ton operating bridge span at a radius of 75 feet.

Initially, several different designs were considered for the wickets. Steel shutters hinged to one side of the wicket frame, for release to swing shut once the frame was in its lowered position, were rejected as impractical. Experiments with sliding timber shutters revealed a serious friction problem. Finally, a facsimile of the Canadian wicket system was adopted with a sliding steel shutter plate or "needle" to be pulled down by a wire cable passing around a sheave in the bottom of the wicket frame.[84]

Both movable bridges, and the wickets, were constructed of structural steel. The 98-foot span "wicket bridge" consisted of two vertical trusses 12 feet high on 12 foot centres. Heavy bracing in the floor, between the lower chords, formed a horizontal truss designed to resist a water pressure of a 28.5-foot head resulting in a horizontal load of 8000 pounds per linear foot. The "operating bridge" of 108-foot span, consisted of two vertical trusses eight feet high on six-foot centres designed to withstand a live load of 15 000 pounds vertical and 10 000 pounds horizontal. These calculations were based on the loads imposed by the weight of a travelling winch truck, that worked the wicket operating cables, and the stress on a cable in lowering each wicket and shutter into a torrent of water under emergency conditions.

The novel derrick-movable emergency wicket dam was erected on the St. Mary's Falls Canal in 1922 to protect the Davis and Sabin locks (Figs. 80 and 81). To test the movable emergency dam, the filling valves of one lock were opened, creating a considerable current. Using the derrick adjacent to the lock, both bridges were placed across the approach channel. The wicket bridge was set down in deep wall recesses, just clear of the water level, and the operating bridge was simply anchored to the wall coping 34 feet upstream. The wicket frames, each three feet wide by just under 34 feet in length, were then swung one after another into position, horizontally, between the two bridge spans. The upper end of a wicket frame was hooked over hinge pins on the upstream truss of the wicket bridge and the bottom end attached to a cable worked by the travelling winch truck on the operating bridge. In this manner, each wicket frame was suspended horizontally between the two bridge spans and then swung down, through the flowing water, to come to rest with its lower end against a timber sill crossing the bottom of the approach channel.

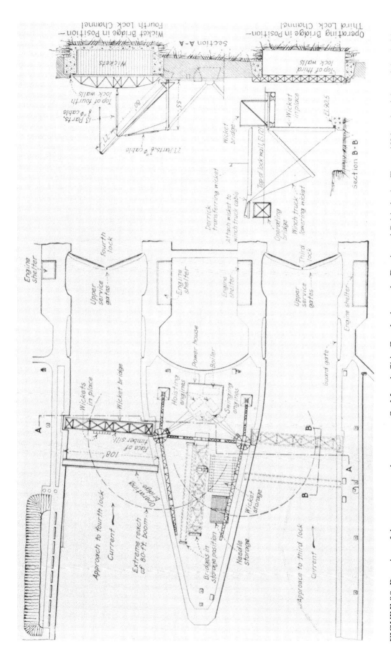

FIGURE 80. Drawing of the emergency dam on new St. Mary's Ship Canal, above the Davis and Sabin locks. Two stiff-leg derricks, wicket bridge, and separate operating bridge are shown in plan and sectional views. (*Engineering News-Record*, 23 October 1924)

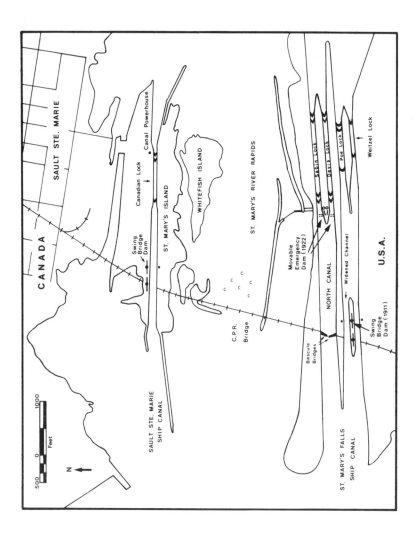

FIGURE 81. Map of the Canadian and American ship canals at Sault Ste. Marie, 1922. (Canadian Parks Service, 1986)

Once all the wicket frames were lowered, the cable for operating the sliding steel shutter in each wicket frame was attached in turn to the travelling winch and the shutters lowered, one after another, to gradually cut off the flow of water. Leakage remained through the gaps between adjacent wicket frames, but the resulting current was judged insufficient to prevent the guard gates from being closed in an actual emergency situation.

The St. Mary's Falls Canal adaptation of the standard emergency swing bridge dam was comparatively cheap to construct — only $250 000 or roughly $1400 per linear foot of opening — including the steam power plant. The wickets were also readily accessible for maintenance and repair on the nose pier, and one wicket dam could be used to close off either canal lock channel.[85] The new derrick-movable emergency wicket dam, however, was not without its drawbacks as canal engineers began to realize following a second installation of this novel adaptation of the conventional emergency swing bridge dam.

The Lake Washington Ship Canal Movable Dam, 1924

The Lake Washington Ship Canal, constructed by the U.S. Army Corps of Engineers in 1911-16, connected Puget Sound on the Pacific Ocean with a large inland freshwater lake — Lake Washington — some 25 feet above sea level at low tide. The eight-mile-long canal consisted of a channel, the prism a minimum of 25 feet deep and 75 feet wide, excavated to connect up several smaller lakes between the Sound and Lake Washington. Two parallel locks were positioned at the lower end of the canal: a large 760 foot by 80 foot ship lock with 36 feet of water on the sills; and a smaller 150 foot by 30 foot lock with 16 feet of water on the sills.

Lake Washington, at the head of the canal, extended over more than 25 000 acres. If the gates of the large lock were destroyed in an accident, a cross-section of water 80 feet wide by 36 feet deep — almost 3000 square feet in area — would be released, flowing at an estimated 58 000 cubic feet per second. Despite the potentially critical situation should the lock gates be destroyed, nothing was done initially to provide protection for either lock. Then the conventional emergency swing bridge dam was investigated. Cost considerations, however, ultimately ruled in favour of the novel derrick-movable emergency wicket dam adaptation developed at the 'Soo'.

The movable emergency dam erected to protect the large ship lock on the Lake Washington Canal consisted of a stiff-leg derrick — 45 feet high, with a boom of 75 tons capacity at a 60-foot radius — a "wicket bridge," and "operating bridge," and detachable wickets. The dam was erected in the same manner as at the 'Soo', but differed in that the derrick and winch on the operating bridge were electrically operated, and the wickets were of an improved design. There were six wicket frames, constructed of steel girders, and 24 shutter plates or wicket gates. Each shutter consisted of a steel-framed buckle plate mounted on roller-bearing wheels, and guided into place on being lowered by a track on the upstream flange of each wicket girder. Four horizontal rows of six shutter plates each, were required to close the emergency dam. As was the case on the Panama Canal, the shutter plates extended across and beyond the face of the wicket frames. Once all the plates were in position across the face of the dam, splines or needles were inserted to close the gaps between the vertical rows of plates (Fig. 82).[86]

The Lake Washington Ship Canal modification of the novel derrick-movable emergency wicket dam was erected in 1924 at a total cost of only $175 000. But it suffered, as did its St. Mary's Falls Canal prototype, from what many canal en-

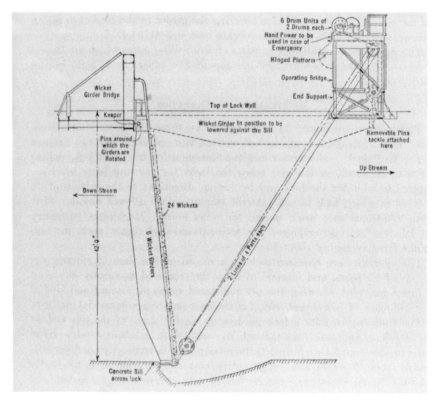

FIGURE 82. Cross-section drawing of the Lake Washington Ship Canal emergency dam, showing movable wicket and operating bridges in operation with the shutter plates in place across the face of the wicket girders. (*American Society of Civil Engineers, Transactions*, 1928)

gineers soon came to consider a serious defect. It took a relatively long time to place this type of dam fully in operation, over four hours in experiments conducted in almost still water on the Lake Washington Ship Canal as opposed to only one hour to operate an emergency swing bridge on the Panama Canal. There was also some concern among canal engineers as to whether the derrick-movable wicket dam, could be assembled without mishap under true emergency conditions.[87]

To close off the smaller lock, with a cross-section 30 feet wide by 16 feet deep, the Seattle engineers adopted a totally different type of movable emergency dam. It consisted of five steel box girders, stored alongside the lock, and a hand-operated stiff-leg derrick. In an emergency, the girders could be swung across the lock channel and lowered horizontally in recesses or grooves in the side walls. Roller wheels, on the ends of each girder, reduced friction while guiding the girders downward in the wall recesses to close off the channel.[88] This was yet another novel type of movable emergency dam, one developed earlier in New Orleans. There canal engineers had adapted, for a new purpose, a far older movable dam structure: the simple stoplog dam, long used to regulate water levels on power canals and as a means of closing off barge canal locks for dewatering to effect repairs.[89]

The New Orleans Stop-Log/Derrick Dam, 1923

At New Orleans a five-mile-long canal, the Inner Navigation Canal, had been built as of 1922 by the U.S. Army Corps of Engineers to connect the lower Mississippi River to a tidal inlet, Lake Pontchartrain, on the Gulf of Mexico. A single lock stepped the canal down 19 feet from the river to the level of the lake. The lock chamber measured 75 feet wide by 55 feet deep — a cross-section of 4125 square feet as compared to the 6017 square feet that had to be closed off by each emergency dam on the Panama Canal. The lay of the land was such that any accident resulting in the destruction of the lock gates would expose a large section of New Orleans to flooding. Such an accident would release a torrent of water with a calculated velocity of 23.2 feet per second, and a volume in excess of 70 000 cubic feet per second — about one-third of the total flow of water over the American and Canadian falls at Niagara.[90]

To protect the city, an emergency dam was designed that consisted of a series of transverse stop logs to be lowered into grooves in the side walls of the lock by a revolving crane positioned on the lock wall. These stop logs were reinforced hollow-box steel girders 84 feet long and six feet deep by almost eight feet wide at their centre. Each weighed 87 tons, and eight of these massive girders were placed horizontally across the lock entrance one on top of another, to close off the canal (Fig. 83).

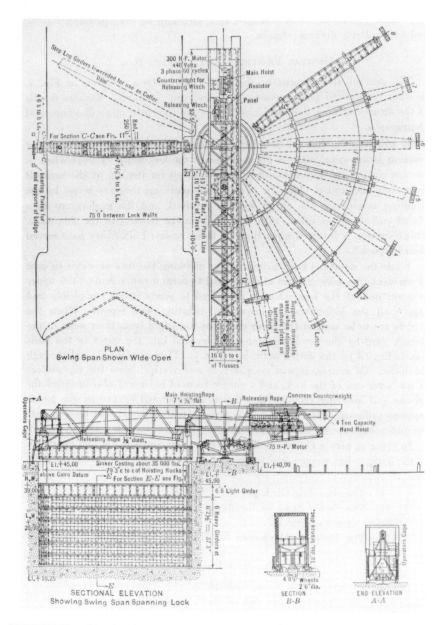

FIGURE 83. Plan of New Orleans Inner Navigation Canal emergency dam, showing movable stop-log/derrick dam in operation and positioning of stop logs on the lock wall coping when not in use. (*American*

The stop-log girders were designed to fill with water on being lowered into the canal to increase their weight, thereby reducing their buoyancy.[91] Massive "sinker castings" — cast iron blocks weighing 17.5 tons each — were also built into the automatic latch releases on the crane hooks to augment the weight of the girder on being lowered into the water.[92] To further help overcome frictional resistance caused by the water pressure acting against the upstream face of a girder on its being slid down in the lock wall grooves, and to ensure a tight fit in the lock wall, vertical steel channels were built into each end to run against continuous bronze guide plates set into the concrete lock wall recess. The stop-log girders were also designed to seat tightly together horizontally, with seals to eliminate any leakage.[93]

When constructed in 1923, it was found that the stop-log girder dam could be closed in about one hour or almost as quickly as the Panama Canal emergency swing bridge dams. The smaller derrick stop-log dam had cost $353 523, but calculated on a comparative basis taking into account changing steel prices, this was less than half the cost per square foot of the Panama dams. Moreover only one man, a crane operator, was required to put the stop log dam in operation, and when tested with a full head of water leakage was almost non-existant — estimated at only 100 to 200 gallons per minute through the horizontal seals between the girders at several isolated spots.[94]

As designed and tested at New Orleans by canal engineers who had worked on the Panama Canal project, the novel emergency stop-log dam proved to have real advantages over the standard emergency swing bridge wicket dam.[95] Indeed, in assessing the design and performance of the stop-log dam, the engineers were for the most part convinced of its overall superiority in all areas of critical concern.[96]

The stop-log dam was not only a relatively cheap and effective means of stemming a torrent of water in an emergency, but was far simpler to design, construct and operate. The revolving crane and grooves in the lock walls negated the need for a large, heavily reinforced, swing bridge superstructure to place and support the movable dam, and stop-log girders appeared far more reliable than wickets. There was comparatively little that could go wrong in operating a stop-log dam; whereas a wicket dam was a complicated machine both structurally and mechanically, known to be susceptible to jamming. Several shutters of the Canadian dam had jammed during the 1909 crisis and on the Panama Canal, wickets occasionally jammed during routine operating drills. Even under extreme emergency conditions, there was little possibility of injury to the heavy rigid girders of the stop log dam. In contrast, there was a concern that the more lightly constructed wicket frame girders might well suffer injury in an emergency. The engineers were well aware

FIGURE 84. Aerial photo, Sault Ste. Marie Canal, 1962. Emergency swing bridge dam and Canadian Pacific Railway swing bridge are alongside the canal just upstream of the lock. (Canadian Parks Service, Sault Ste. Marie Canal Office)

that one of the wicket frames of the Canadian dam had been badly twisted, apparently by the strong current of water striking it from an angle.

Where repairs and maintenance were concerned, the stop-log dam also appeared superior. Leakage was infinitesimal, eliminating the need for employing a caisson or guard gates to close off the canal for dewatering and repairs, and there were few working parts to get out of order. Moreover, all components were readily accessible for inspection, routine maintenance and repairs, a critical feature on taking human factors into account.

It was realized that with emergency structures resting unused for long periods of time, there was a natural tendency to neglect routine maintenance and inspections with the result that the machinery might not be fully operational when put in service during an emergency. On the Panama Canal, the swing bridge dams were operated once a month as a matter of routine to overcome this potential problem. Nonetheless, the fact remained that on wicket dams with much of their machinery relatively inaccessible under the swing bridge superstructure, this tendency was likely to be all the more pronounced and ultimately dangerous.[97]

Following the development of the novel emergency stop-log dam at New Orleans, canal engineers turned away from the older swing bridge wicket dam concept. The Panama Canal structures marked the highest development of the emergency swing bridge wicket dam, but were the last of the Sault Ste. Marie prototype ever constructed.[98] Thereafter on major ship canal construction projects where a movable emergency dam was required — such as at the summit level of the Fourth Welland Canal, completed in 1932 as an improvement on the earlier ship canal by-passing Niagara Falls — some variation of the stop-log dam system was invariably used.[99] Indeed, eventually, all but one of the existing emergency swing bridge dams were removed from service.

On the Panama Canal, all six emergency swing bridge dams were dismantled and sold for scrap: three in 1955; and the remaining three several years later. Having never been used in an an emergency over a period of some 40 years, they had come to be considered superfluous with the other safety features — the towing locomotives, fender chains, and heavy steel guard gates — in place.[100] At Sault Ste. Marie, the American swing bridge dam of 1910-11 was replaced by a derrick/stop-log system on the building of the new MacArthur lock in 1942-43.[101] Only the Canadian swing bridge dam on the 'Soo' Canal remains in service for emergency use, and it has suffered maintenance problems associated with that type of structure (Fig. 84).

"the temptation is to neglect proper inspection and upkeep"

Following its rehabilitation in 1910-11, little attention appears to have been paid to the Canadian emergency swing bridge dam at Sault Ste. Marie. The structure was scraped and painted at roughly 10-year intervals over the following decades, and minimal maintenance work was required. During the winter of 1929-30, the bevelled timber seals between the wicket frames were renewed and creosoted. In 1939-40, part of the wood flooring on the bridge was renewed and a new end rest girder erected under the long arm of the swing bridge superstructure.[102]

Over the course of three decades, the swing bridge dam was fully operated only once. From December 1933 through to February 1934, the swing bridge dam was used as a coffer dam — no doubt in conjunction with the upper guard gates — while repair work was carried out on the southwest wall of the lock (Fig. 65).[103] In the other years, an effort was made to swing the bridge, but only after the close of the navigation season to avoid any delay to shipping. This winter exercise consisted simply of swinging the bridge across the canal and lowering one or two wickets onto the ice (Figs. 85 and 86).

The wickets were lowered and raised manually using the chain hoist. Steam was got up in the boiler, but used only for melting the accumulation of ice on the turntable track and the turning gear rack to facilitate the swinging of the bridge.[104]

FIGURE 85. Closing the emergency swing bridge dam across the canal channel, prior to lowering the wickets onto the ice during a winter inspection. (R. Draycott, Sault Ste. Marie Canal Office, 1983)

FIGURE 86. Deck of the emergency swing bridge dam, Sault Ste. Marie Canal. Hoist chain hangs at right, with wicket chains on deck locked in a chain stop to suspend the bottom end of the wicket frame up under the bridge floor. (R. Draycott, Sault Ste. Marie Canal Office, January 1983)

During the Second World War as vast quantities of iron ore were shipped to steel mills of the East, both 'Soo' canals came to have the utmost strategic importance. The Canadian government became concerned about preventing acts of sabotage and keeping the Canadian canal in uninterrupted operation. To that end, troops were stationed at the canal site,[105] and the Department of Transport despatched the Assistant Engineer of the Welland Canal, J.B. McAndrew, to examine and report on the mechanical and structural equipment of the Canadian 'Soo' canal, including the emergency swing bridge dam.

McAndrew found the swing bridge dam to be structurally sound; yet in need of maintenance work. The steam hoist boiler was in good shape, and the machinery of the wickets well greased. But several roller wheels were missing from the wicket frames, and there was no sign of any lubricant on the turntable.[106] Indeed, the turntable had not been greased for some time, and the results were obvious. The previous year, eight men had been needed to swing the bridge, and occasionally in the past a truck or tug had had to be used to start the bridge in motion.[107] In contrast, when newly erected, two men had easily swung it with the turning lever.[108]

In his report of 24 November 1942, McAndrew recommended that all working parts of the swing bridge dam be thoroughly lubricated and a maintenance program established to ensure that the bridge would be swung, and all the wickets fully lowered, at least once a year during the navigation season. Moreover, he asserted that the hand-operated turntable of the bridge, the end lift latches on the rest girder, and the wicket chain hoist ought to be converted to electrical operation. The steam hoist, in particular, should be removed and replaced by a 20-horsepower winch, specifically one of the electrically powered boat-hauling winches that were no longer needed at the Prescott grain elevators on the St. Lawrence River.[109]

During the winter of 1942-43, detailed plans were prepared by McAndrew for converting the emergency swing bridge dam to electrical operation, but the winch proposal proved abortive. McAndrew had proposed to simply substitute an electric winch for the steam hoist, making use of the wire rope, blocks and tackles of the steam powered hoist system. But on a closer examination too many alterations were required to adapt the existing system to an electric winch.[110] The canal engineers, however, remained determined to convert the wickets to electrical power.

Where the operation of the wickets was concerned, neither the manually operated chain-fall hoist nor the steam hoist had proved completely satisfactory. With the chain-fall system, the chain block hook was inserted into a link in the wicket chain at deck level just above the chain stop. The chain was taken up to release the chain stop, and then let out to swing down the wicket. The chain-fall, however, had a limited "grab" or travel. In lowering a wicket, the chain had to be stopped at 11-foot intervals and inserted again into the chain stop. The chain block hook then was released, and taken back up to repeat the process letting out another

11 feet of chain, and so on, until the wicket frame seated itself against the underwater sill and the shutter slid down to close the wicket. At that point, the wicket chain was locked again with the chain stop, and the chain-fall hook removed (Figs. 66 and 86).

In closing the wickets, the chain hoist was moved along the overhead trolley track from one wicket to the next, skipping over every other wicket, and then back again along the trolley track to lower the alternate wicket frames in between. This was a frustratingly slow system as there was 60 feet of chain on each of the 23 wickets, with over 30 feet of chain travel being required to complete a full wicket/shutter plate closing, or opening, sequence.

With the steam hoist system, the wicket chain could be let out without stopping during the wicket closing operation, but it took about two hours to get steam up in the boiler before work could commence. Moreover in the steam hoist system, the wire rope cable and its blocks and tackles had to be moved from one wicket to another, a slow and cumbersome operation.

Thwarted in his electric winch proposal, McAndrew proceeded to adapt the existing chain-fall system to electrical power. He planned simply to substitute an electrically powered chain block for the manually operated chain fall on the overhead trolley beam. In this system, the wicket chain would be put directly onto the chain sprocket in the hoist, and the chain run through until the slack was taken up and the load lifted off the chain stop. Then the chain could be let out without stopping during both phases of the wicket closing operation. At the completion of each phase, the wicket chain would be locked again in the chain stop. The rest of the chain had simply to be run out of the chain block, and the hoist moved along the trolley track to the next wicket where the process would be repeated.[111]

According to McAndrew's plans, an electrically powered chain block could easily be mounted on the existing trolley with three-phase a.c. receptacles positioned along the bridge superstructure for plugging in the chord of the hoist motor on its being moved along the trolley track. The standard chain block of the power required operated at a speed of four feet per minute, but this was to be increased to 12 feet per minute to speed the wicket closing.[112] Plans were also prepared for employing a 15-horsepower motor to swing the bridge, and McAndrew recommended that the cast iron turning gears and shafts be replaced with cut steel gears.[113] Nothing, however, came of any of these plans.

The Department of Transport was not convinced of the need for further expenditures on the emergency swing bridge dam. The steam hoist system appeared to be adequate to get the job done.[114] Moreover where the swinging of the bridge was concerned, electrification was far from a necessity. Even during the period when the turntable gears had been left unlubricated, an eight-man force had still managed to manually swing the bridge closed in under three minutes.[115]

In the immediate post-war years, the tonnage of freight passing through both 'Soo' canals continued to increase over the high wartime levels,[116] but the maintenance and operational concerns raised by McAndrew were forgotten. In 1947, the canal superintendent's office reported that the swing bridge dam was closed across the canal "nearly every year for practice," but the wickets were not being lowered. Nonetheless, the swing bridge dam appeared in good condition and when last used in 1934, no difficulty had been experienced with the wickets.[117] Such operational and maintenance practices, however, were soon found to be grossly inadequate.

"... so badly eroded with rust"

As late as July 1950, all appeared well with the emergency swing bridge dam. In that month, several bevelled timber seals on the wicket frames were renewed, all working parts were thoroughly lubricated, and the steam hoist and boiler were tested during an inspection. The canal superintendent reported that: "The dam was swung over the canal as far as possible and found to be in very good working condition."[118] But appearances were deceiving. Within a month, the Department of Transport was informed that the swing bridge dam was inoperable and in need of a drastic overhaul.

The Department reacted quickly. An engineer, E.C. Shurly, was despatched on the evening train from Ottawa to determine the extent of the repairs required on the 'Soo' Canal, and Dominion Bridge was contacted to send a representative empowered to give "a firm price" on the spot for the work.[119] On arrival, Shurly had the canal force shift four of the sliding shutter plates to the bottom of their wicket frames and lower the wickets onto the ground for inspection.

Shurly spent two days examining the wickets, the second day in the company of two Dominion Bridge representatives, and on returning to Ottawa reported that:

> the four bottom buckle plates of the wickets were so badly eroded with rust as to be absolutely unsafe, in fact on several of these large areas had been eaten completely through....
>
> The bottom rollers on which the [shutters] slide into position in the frames were rusted in a solid mass to their supporting angles, and these in turn were so badly rusted as to require replacement.[120]

The four bottom buckle plates in all 23 shutters would probably have to be replaced, as well as the flange angles supporting the roller wheels, all of the roller wheel pins, and six missing roller wheels. One shutter plate with a badly bent frame had to be straightened to render it operational, and approximately 50 percent of the cross-bracing angles in the bottom panel of the wicket frames needed to be cut out and replaced. These repairs would demand about 40 tons of new material, and require about eight weeks of fieldwork. On a cost plus basis, the job was estimated at about $35 000 to $40 000.[121]

With the assessment report and preliminary estimate in hand, C.W. West, the Director of Canal Services, immediately notified the Deputy Minister of the crisis situation, and addressed a memorandum to the Administration and Legal Services to obtain approval for transferring monies freed up in the Rideau and Trent canal budgets to cover the unexpected expenditure at Sault Ste. Marie. West wrote:

> ... recent investigations by canal engineers show that although maintenance has been kept up, certain important sections of the [shutter plate] valves and [wicket]frames, not readily accessible, have rusted out and should be replaced.
> ... the structure is of a vital importance ..., it should be placed in good working condition as soon as possible.

This would require the transfer of $43 000 to the 'Soo' Canal, a sum that included $3000 to purchase and install an electric hoist to improve the wicket operating system.[122]

Once the transfer of monies was approved, specifications were prepared as of 14 December 1950, and with the canal closed for the winter, work proceeded in the new year on the rehabilitation of the swing bridge dam wickets. The specifications covered all of the work identified earlier, as well as provided for the removal, cleaning, and re-installing of all the roller wheels, and more generally, the repair or replacement of all parts found to be badly rusted. All of the work was to be completed by 31 March 1951, in time for the opening of the next year's navigation.[123] When the wickets were removed from under the swing bridge superstructure, however, rust damage was found to be much more severe than anticipated. This necessitated a good deal of extra work, and resulted in several design modifications.

Where the wicket frames were concerned, the main problem was the roller pins. Through long disuse, the pins had become so badly seized in the cast iron roller wheels and wicket frame journals as to defy removal.[124] Hence new roller wheels, as well as new pins, were required, and the rollers were to be equipped with Oilite bronze bearings to minimize any tendency to freeze through long disuse.[125] As a further precaution, it was decided to replace all of the hinge pins of the wickets in the existing cast steel hinge brackets, and the wicket chains were sent to Dominion Bridge's Montreal plant to be annealed and have any damaged links or shackles renewed.[126] The shutter plates of the wickets, however, presented far greater problems.

On removing the sliding shutter plates from the wicket frames, it was discovered that many of the I-beams forming the sides of the structural framework were badly rusted. This was particularly the case along the bottom half of the frames where the strength of many I-beams was completely destroyed by corrosion. In one instance, there were rust holes right through the web of a beam. The side beams of all 23 shutter plates had to be replaced.[127] Finding an immediate supply for such a large quantity of 7.125-inch I-beams, proved impossible. A number were found in stock at Montreal, but nowhere else in Canada or the United States. Finally, anagreement was made with United States Steel to give priority in rolling to the I-beams for delivery a month later, in March 1951.[128]

In rebuilding the shutter plate frames, web plates were replaced where required at the cross-members, and both welding as well as rivetting was used in placing the buckle plates back on the frame panels, depending on the condition of the plates.[129] Efforts to replace the rusted out buckle plates in the bottom four panels of each shutter brought even further difficulties.

During the rehabilitation of the emergency swing bridge dam, the Sault Structural Steel Company supplied most of the necessary steel shapes, such as the angles for supporting the roller wheels, and was responsible for the fieldwork. Its parent company, the Dominion Bridge Company of Montreal supplied the pins and roller wheels, some of the shutter frame I-beams, and had planned to provide the buckle plates.[130] But the 1/4-inch steel plates could not be found in stock anywhere in North America, or obtained on short notice even if given a priority in rolling. Hence, arrangements were made to have 1/4-inch plate shipped from stock in Britain with a projected 1 April delivery date.[131]

To avoid further delays and additional costs, a design change was adopted. Rather than buckling or "dishing" the plates at Montreal to duplicate the old buckle plates, Dominion Bridge suggested that a six-inch channel stiffener be placed across the centre of each of the bottom four panels in the shutter frame. With the support of these extra stiffeners, the 1/4-inch plate could be welded on flat over the full length of the bottom four panels. This suggestion was approved and the flat plates, which arrived from Britain on 31 March, were shipped directly from Montreal to the 'Soo' for installation.[132]

As of 26 April, all of the welding and rivetting work was completed, and the shutters were slid back into the wicket frames.[133] On re-assembling the wickets, one further change was made. Bolts, rather than rivets, were used in putting back the diaphragm or strut across the bottom of each wicket frame. The shutter plates, hitherto inaccessible in the wicket frames, could now be pulled out for future inspections.[134] Four days later, on the opening of the 1951 navigation season, the emergency swing bridge dam was fully operational and all concerned were satisfied with the rapid completion of the project.[135]

With the extra work undertaken during the overhaul of the swing bridge dam wickets, the final cost reached $54 007.96 an increase of 35 percent over the preliminary estimate.[136] To meet this immediate demand, funds were re-allocated from an on-going construction project on the Lachine Canal, and ultimately a supplementary estimate was submitted to Parliament.[137] All that remained was to electrify the wicket operating system and paint the structure.

Arrangements had been made to obtain a 7.5-ton electric hoist from Provincial Engineering Ltd. of Niagara Falls, Ontario. This three-horsepower hoist had an operating speed of four feet per minute, and was to be suspended from the shackle on the existing trolley and track.[138] But this proved impossible. The hoist required a 15-inch clearance between the trolley track and the bridge superstructure. At some points, clearance was less than 11 inches and on several sections the trolley track was found badly bent.[139] To resolve this problem a new trolley track and trolley, suitable for both the electric hoist and the steam hoist auxiliary system, was designed and fabricated.[140] The new hoist system was finally in place as of March 1953, completing the rehabilitation and electrification of the swing bridge dam wickets (Fig. 87).[141]

Three years later, the Department ordered a thorough testing and report on the rehabilitated swing bridge dam.

Operating Test: December 1956

At the close of navigation in mid-December 1956, a large workforce was assembled to operate the emergency swing bridge dam. The removal of accumulated ice and snow caused some delay, with 35 minutes being required to release the end lifts, swing the bridge, and secure it against the abutment stops. Then work commenced in lowering the wickets in still water.

Initially, the electric hoist was used. But it proved very slow, taking up to 36 minutes to lower and close a single wicket. As soon as steam pressure was up in the boiler, after three wickets were closed and a fourth half-closed, a switch was made to the steam hoist. The remaining 19 and a half wickets were closed in two hours and 30 minutes, an average of eight minutes and 28 seconds per wicket. In effect, the steam hoist worked four to five times as fast as the electric hoist. However, a 10-man crew was required to operate the steam hoist as opposed to only four men on the electric hoist.

During the test, the upper valves of the lock chamber just downstream were opened to create a current prior to the lowering of the last two wickets. They were closed in the flowing water without any difficulty. As a further test, the lock was alternately emptied and filled to lower the water level in the reach below the swing

FIGURE 87. Electric hoist and trolley track installed on the emergency swing bridge dam in 1953. Wick-et chain is being raised, to release it from the chain stop, preparatory to lowering a wicket for inspection. (R. Draycott, Sault Ste. Marie Canal Office, January 1983)

bridge dam, placing the structure under the strain of a full head of water. At that point, one of the timber stop blocks compressed slightly. The outer end of the bridge swung three inches forward, but no damage was done.

The ensuing report made several recommendations. Once a month during the navigation season, the swing bridge ought to be swung and the hoisting equipment thoroughly inspected. The boiler ought to be inspected annually, and the bridge swung each winter and all wickets lowered onto the ice for examination. The electric hoist also needed to be speeded up, and a booklet prepared to instruct the workforce in, "How to operate the movable dam."[142]

Although no mention was made of McAndrew's earlier hoist plan, the problems with the existing electric hoist were obvious. The speed of the standard electric hoist had not been increased from four to 12 feet per minute as McAndrew had intended back in 1942-43, and the electric hoist chain had a very limited "grab" or travel. Unlike McAndrew's proposed electric chain block, the wicket chain did not feed directly into the sprocket of the new hoist. In effect, the electric hoist had the same deficiency as the earlier manually operated chain-fall hoist. The wicket chain could not be let out in one continuous operation, as it was with the steam hoist. To the contrary, the wicket chain had to be let out in short lengths corresponding to the length of the hoist chain, and locked in the chain stop at the end of each travel while the hoist chain was run back to take another grab.

The test report was submitted to the Department of Transport in January 1957, but no effort was made to modify the electric hoist. The steam hoist, however, was improved. A smaller, more efficient, portable boiler was purchased, superseding the old vertical boiler on the swing bridge dam.[143]

Maintaining the Structure: (1959-85)

In April 1959, responsibility for operating and maintaining the Sault Ste. Marie Canal was transferred from the Department of Transportation to the new St. Lawrence Seaway Authority.[144] Under the Seaway Authority, the emergency swing bridge dam underwent one major rehabilitation. During the winter of 1972-73, the concrete counterweight on the short arm of the swing bridge superstructure was renewed, including the floor beams and stringers under the counterweight. At the same time the bevelled timber seals on the wicket frames, the roller wheels in the wicket frames, and the wooden deck on the swing bridge were replaced. When completed, the whole swing bridge dam was chipped and painted, bringing the total cost of the rehabilitation project to $92 000.[145]

Thoroughly rehabilitated, the emergency swing bridge dam continued to play a critical role in standing by to protect the Sault Ste. Marie Canal should the lock gates be severely damaged or carried away in an accident. But by the late 1970s, the prospect of such an occurrence appeared rather remote. The large lake freighters had long since totally abandoned the Sault Ste. Marie Canal in favour of the deeper draught/larger locks on the American canal system, and the passenger steamship era had drawn to a close. Small pleasure boats and tour boats constituted the only traffic through the Canadian canal.

No longer of any commercial importance as a ship navigation, the Sault Ste. Marie Canal and adjacent federal lands were transferred from the St. Lawrence Seaway Authority to Parks Canada (now the Canadian Parks Service, Environment

Canada) in April 1979.[146] Henceforth, Parks was responsible for operating and maintaining the canal with a mandate to preserve and interpret its heritage character and resources.[147] The emergency swing bridge dam, in being an integral part of the canal, came under this mandate.

Under the aegis of Parks, a fourth rehabilitation of the emergency swing bridge dam was undertaken during the summer of 1985. All of the wickets were disassembled, the various components examined for soundness, repaired or replaced where required, and sand-blasted prior to being painted and re-assembled (Figs. 88 and 89). The swing bridge superstructure and turntable gearing were also sandblasted and painted. No design modifications, however, were carried out on the swing bridge structure.[148]

FIGURE 88. Wicket frame of the emergency swing bridge dam, with the sliding shutter removed during rehabilitation work. (R.W. Passfield, Canadian Parks Service, July 1985)

The Sole Survivor

The newly refurbished emergency swing bridge dam on the Canadian 'Soo' Canal is the sole survivor of a novel type of structure designed to afford protection to ship canals stepping up into vast bodies of water. Although the danger of a serious accident at Sault Ste. Marie is now minimal, the canal having long since passed its heyday as one of the world's busiest ship navigations, the swing bridge dam still fulfills its original function. It stands ready for use in an emergency to close off the canal channel and stem any torrent of water unleashed should the lock gates be badly damaged or carried away.

This emergency swing bridge dam is the second of its type ever constructed. Although based on a design concept and principle of operation developed earlier on the American St. Mary's Falls Canal at Sault Ste. Marie, Michigan, the Canadian structure introduced several innovations in its wicket design and operation that became a standard feature on subsequent emergency swing bridge dams. It also introduced a new building material, steel, to swing bridge dam construction in place of the iron and wood used in its sole predecessor.

The Canadian structure is the only swing bridge dam to be operated under extreme emergency conditions and, during the June 1909 crisis, proved the practicability of a hitherto untested structure to a skeptical engineering profession. The Sault Ste. Marie Canal emergency swing bridge dam also served as the design prototype for improved structures erected subsequently on an enlarged St. Mary's Falls Canal in 1910-11, and on the Panama Canal in 1914.

FIGURE 89. Rehabilitating the emergency swing bridge dam, Sault Ste. Marie Ship Canal. Workmen place newly restored wickets back beneath the swing bridge superstructure. (R.W. Passfield, Canadian Parks Service, July 1985)

For almost three decades following its erection in 1895, the Sault Ste. Marie swing bridge dam was regarded as the standard, and for a time the only practicable and proven device for protecting major ship canals threatened by potentially raging torrents of water. But only nine emergency swing bridge dams were ever constructed. The increasing cost and complexity of swing bridge dams resulted as of the 1920s in their being superseded by a new derrick-stop log emergency dam that proved far less costly, less complicated both mechanically and structurally, and much easier to maintain and operate. Indeed, maintenance proved a costly on-going problem with emergency swing bridge dams as illustrated by the subsequent history of the Canadian structure.

The emergency swing bridge dam at Sault Ste. Marie has undergone four major rehabilitations during its 90-year existence, but remains essentially as constructed in 1895. Much of the steelwork of the wicket frames and shutters has been replaced in kind, and only one significant design modification has been made — the addition, in 1911, of bevelled timber seals to close the gap between the wicket frames. Several minor modifications have been made over the years to improve the ease of operation through adding a steam hoist, and subsequently substituting an electric hoist for the original manually operated chain hoist. But the operating principle, and the structural and mechanical integrity of the swing bridge dam, has remained unimpaired.

The Canadian emergency swing bridge dam is an unique structure, outstanding in its innovative wicket design, its operating history, and its continuous long-term service — truly one of a kind.

Endnotes

1   Occasionally guard gates were constructed so as to form a second lock chamber — a so-called "guard lock," with no lift — just above and in conjunction with the main lock. See Robert W. Passfield, *Building the Rideau Canal: A Pictorial History* (Toronto: Fitzhenry & Whiteside, 1982), p. 66.

2   Henry Goldmark, "Emergency Dam on Inner Navigation Canal at New Orleans, Louisiana," *American Society of Civil Engineers* [hereafter cited as *ASCE*], *Transactions*, Paper No. 1695, Vol. 92 (1928), p. 1590.

3   "Accident to the Sault Ste. Marie Canal Lock," *The Engineering and Mining Journal* (30 August 1890), p. 241.

4   "The New United States Government Lock at Sault Ste. Marie, Michigan," *Engineering News* (16 September 1895), p. 194.

   These original Sault Ste. Marie locks were also of a stupendous size. As of mid-century, the newly completed Second Welland Canal had locks 150 feet by 26 feet six inches with nine feet, soon increased to 10 feet, of water on the sills; and the St. Lawrence canals had locks 200 feet by 45 feet with nine feet of water on the sills. In Britain, the largest locks, 180 feet by 40 feet, were on Telford's 20-foot-deep Caledonian Ship Canal navigation (1804-22), bisecting Scotland. The Erie Barge Canal was about to be enlarged, ca. 1854-62, from locks 90 feet by 15 feet with four feet of water on the sills to locks 110 feet by 18 feet with seven feet of water on the sills.

5   W.P. Kibbee, "The Busiest Canal in the World," *Engineering Magazine* (July 1897), p. 601.

6   "The New United States Government Lock at Sault Ste. Marie, Michigan," *Engineering News* (16 September 1895), p. 194; and "Accident to the Sault Ste. Marie Canal Lock," *The Engineering and Mining Journal* (30 August 1890), pp. 241-242. Guard gates were constructed at either end of the large lock to facilitate dewatering when repairs were required.

   Barge canal locks generally had lifts of anywhere from six to eight feet, and occasionally 10 feet. Ship canals ca. 1880 had similar lifts. Higher lifts, however, were not unknown. Four locks constructed at Jones' Falls on the Rideau Canal in 1828-32, had lifts of 15 feet each — a stupendous lift for canal locks at that time.

7   Goldmark, "Emergency Dam," *ASCE, Transactions*, p. 1591.

   Alfred Noble was assistant engineer, 1873-83, under Major Godfrey Weitzel during the construction of the Weitzel lock (see "Noble, Alfred," *Dictionary of American Biography*, Vol. 8 [1934], pp. 536-537).

8   "St. Mary's Falls Canal Bridge and Dam," *Engineering and Building Record* and *Sanitary Engineer* [hereafter cited as *Engineering Record*] (22 March 1890), pp. 246-248; and United States Army, *Annual Report of the Chief of Engineers, United States Army to the Secretary of War for the Year 1881*, Washington: Government Printing Office, 1881, Part III, Appendix JJ1, "Improvement of Saint Mary's Falls Canal," p. 2267.

9   Goldmark, "Emergency Dam," *ASCE, Transactions*, p. 1591.

10   Canada, Department of Railways and Canals [hereafter cited as DRC], *Annual Report*, 1894, p. lxxvii.

     Elsewhere by the 1890s, ship canals were being constructed with locks that exceeded the 515 foot by 80 foot Weitzel lock colossus of 1881, but did not attain the length of the 900 foot by 60 foot Canadian lock of 1895. In Britain, the Manchester Ship Canal (built 1887-94) had five pairs of locks as large as 600 feet by 80 feet and lifts from 13 feet to 16 feet six inches on a 26-foot-deep navigation 35.5 miles long. In Germany, the Kiel Canal (1887-1895) had a pair of tidal locks 720 feet by 82 feet at each end of a 61.5-mile-long sea-level navigation, 24.4 feet deep.

11   "Accident at the Canadian Lock at Sault Ste. Marie," *Engineering Record* (19 June 1909), p. 792; and Goldmark, "Emergency Dam," *ASCE, Transactions*, p. 1630.

12   "Accident at the Canadian Lock at Sault Ste. Marie," *Engineering Record* (19 June 1909), p. 792.

13   "St. Mary's Falls Canal Bridge and Dam," *Engineering Record* (22 March 1890), p. 246.

14   "The Panama Canal - No. V," *Engineering, An Illustrated Weekly Journal* [hereafter cited as *Engineering* (London)] (18 July 1913), p. 71.

15   Canal Records, File C-4256/S32-10, Vol. 1, "Sault Ste. Marie Canal, Specification, For a Movable Dam and Swing Bridge of Steel," Robert C. Douglas, Hydraulic and Bridge Engineer, DRC, 15 January 1895. The wicket dam was to be capable of withstanding a 26-foot head of static water on the canal being dewatered for effecting repairs, and capable of being operated in a current generated by an 18-foot drop in water level should the lock gates be destroyed.

16   Ibid., Dominion Bridge Company, "Sault Ste. Marie Canal, Tender for furnishing materials, workmanship, and the construction and erection of a Movable Dam and Swing Bridge over the Sault Ste. Marie Canal," 14 February 1895. The company submitted three alternative designs (ibid.). The approved design was "drawing no. 6, marked D" (ibid., Articles of Agreement, Railways and Canals and The Dominion Bridge Company, 27 March 1895). See

Sault Ste. Marie Canal Office, Drawing No. 50464, Case D, "Soo Canal, Movable Dam, Plan D, Alternative Design for Wickets & Frames."

17   "The Movable Dam and Swing Bridge on the Sault Ste. Marie Canal," *The Canadian Engineer*, Vol. 4, No. 4 (August 1896), p. 98. The improvements suggested by Dominion Bridge actually lowered the cost of the structure. Dominion Bridge had bid $84 000 to construct the emergency dam as originally designed and the other bidder, the Central Bridge Engineering Company of Peterborough, had bid $88 000 (Canal Records, File C-4256/S32-10, Vol. 1, "Sault Ste. Marie Canal, Abstract of Tenders received for Movable Dam & Swing Bridge," 15 February 1895).

18   Canal Records, ibid., William Crawford to C. Schreiber, 23 August 1895; and ibid., W. Crawford to C. Schreiber, 26 October 1895.

19   Ibid., "Sault Ste. Marie Canal, Specification, For a Movable Dam and Swing Bridge of Steel," 15 January 1895; and The St. Lawrence Seaway Authority, Sault Ste. Marie Canal, Case D, Drawing No. 50344, "Movable Dam and Support Bridge over Canal," J.D. Bouchard, 14 March 1968.

20   Ibid., "Sault Ste. Marie Canal, Specification, For a Movable Dam and Swing Bridge of Steel," 15 January 1895.

21   "The Movable Dam and Swing Bridge on the Sault Ste. Marie Canal," *The Canadian Engineer*, August 1896, p. 98.

22   See "The Accident to the Lock Gates at the Canadian Canal, Sault Ste. Marie," *Engineering News, A Journal of Civil, Mechanical, Mining and Electrical Engineering* [hereafter cited as *Engineering News, A Journal*] (17 June 1909), p. 673; Canal Records, File C-4256/S32-10, Vol. 1, "Soo Canal - Movable Dam, Plan, Alternate Design for Wickets and Frames," 7 February 1895; Sault Ste. Marie Canal Office, Drawing No. 50344, Case D, "Movable Dam & Support Bridge over Canal," 14 March 1968; and ibid., Drawing No. 50464, Case D, "Soo Canal, Movable Dam, Plan D, Alternative Design for Wickets and Frames," Dominion Bridge Company, n.d. (February 1895).

23   Ibid.; and Canal Records, File C-4256/S32-10, Vol. 3, C.W. West to E.B. Jost, 13 August 1943, letter describing the chain hoist operation.

The chain-fall hoist was also a Dominion Bridge Company innovation. The original specifications called for a crab windlass or crabs to operate two wicket frames simultaneously from each end of the bridge. Each crab windlass was to be capable of exerting a pull of 30 000 pounds and of holding a stress of 50 000 pounds. The crabs were also to be double geared, capable of being operated at a slow speed by two men and at a faster speed by four men, and equipped with safety pawls and handbrakes (ibid., Vol. 1,

Specification, for a Movable Dam and Swing Bridge of Steel, 15 January 1895).

24 Canal Records, File C-4256/S32-10, Vol. 1, Crawford to Schreiber, 23 January 1894, enclosed drawings: "Plan of Foundation for proposed Movable Dam"; and "Plan of Masonry Platform of Movable Dam."

25 DRC, *Annual Report*, 1896, p. 97 and 1897, p. 119; Canal Records, File C-4256/S32-10, Vol. 1, V. Curran to C. Schreiber, 4 May 1896; and ibid., Robert C. Douglas, Memorandum on Swing Dam, 12 May 1896.

In designing the substructure of the emergency swing bridge dam, the plan of the earlier American swing bridge dam was followed with but one exception. Portland cement concrete was to be substituted for the timber and natural cement foundation constructed to support the underwater sill of the American swing bridge dam (ibid., Crawford to Schreiber, 23 January 1894). But the remainder of the Canadian swing bridge dam substructure was still to be of stone masonry construction as of that date.

26 Ibid., Vol. 1, J. Boyd, Superintendent, Sault Ste. Marie Canal, to C. Schreiber, Chief Engineer, DRC, 29 April 1896; and ibid., Robert C. Douglas to C. Schreiber, 30 April 1896.

27 "The Movable Dam and Swing Bridge on the Sault Ste. Marie Canal," *The Canadian Engineer* (August 1896), p. 98.

28 Ibid.; and Canal Records, File C-4250/S32-10, Vol. 1, "Sault Ste. Marie Canal, Specification for a Movable Dam and Swing Bridge of Steel," 15 January 1895.

29 "The Movable Dam and Swing Bridge on the Sault Ste. Marie Canal," *The Canadian Engineer* (August 1896), p. 98.

30 "The St. Mary's Falls Canal Bridge and Dam," *Engineering Record* (22 March 1890), pp. 246-248, and Figs. 3, 11, and 15.

31 Goldmark, "Emergency Dam," *ASCE, Transactions*, p. 1630.

32 DRC, *Annual Report*, 1901, p. 210.

33 "The Accident at the Canadian Lock at Sault Ste. Marie," *Engineering Record* (19 June 1909), p. 789.

34 Ibid. The *Assiniboia* was 346 feet long with a 44-foot beam, and the *Crescent City* 424 feet by 48 feet, and they could be locked through the 900 foot by 60 foot lock one astern of another. The Canadian lock was designed to take three freighters in line, but ca. 1909 this was no longer possible owing to the increased size of lake freighters.

35 Ibid., pp. 789-790; and DRC, *Annual Report*, 1910, pp. 263-264. All three ships were damaged, by collisions with each other or the lock walls. The *Assiniboia*, however, proceeded directly to Owen Sound. The *Percy G. Walker*

and the *Crescent City* were towed to the American side and repaired in a few days.

36   Ibid. The lower guard gates were used only when dewatering the lock chamber.

37   "The Accident at the Canadian Lock at Sault Ste. Marie," *Engineering Record* (19 June 1909), p. 790; and Canal Records, File C-4256/S32-10, Vol. 2, J.W. Le B. Ross, Superintendent, Sault Ste. Marie Canal, Memorandum, 6 July 1909.

38   "The Accident at the Canadian Lock at Sault Ste. Marie," *Engineering Record* (19 June 1909), p. 791; and DRC, *Annual Report*, 1910, p. 264. See also Osborne and Swainson, *The Sault Ste. Marie Canal, A Chapter in the History of Great Lakes Transport*, pp. 87-92.

39   DRC, *Annual Report*, 1910, p. 264; "The Accident at the Canadian Lock at Sault Ste. Marie," *Engineering Record* (19 June 1909), pp. 790-791; and Canal Records, File C-4256/S32-10, Vol. 2, J.W. Le B. Ross, Memorandum, 6 July 1909.

40   DRC, *Annual Report*, 1910, p. 264; and "The Accident at the Canadian Lock at Sault Ste. Marie," *Engineering Record* (19 June 1909), p. 792. On the mitre posts of the two gate leaves coming together against the sill, the deformation was such that the tops of the gates were still four feet apart.

41   DRC, *Annual Report*, 1910, p. 264.

42   "The Accident at the Canadian Lock at Sault Ste. Marie," *Engineering Record* (19 June 1909), p. 792; "The Accident to the Lock Gates at the Canadian Canal, Sault Ste. Marie," *Engineering News, A Journal* (17 June 1909), p. 673; and "May Re-open Today, Lock Nearly Ready for Business Again," *The Sault Star*, 17 June 1909.

43   DRC, *Annual Report*, 1910, p. 265; and "The Accident at the Canadian Lock at Sault Ste. Marie," *Engineering Record* (19 June 1909), p. 792.

44   Ibid., p. 789; and Goldmark, "Emergency Dam," *ASCE, Transactions*, pp. 1622 and 1635.

45   "The Accident at the Canadian Lock at Sault Ste. Marie," *Engineering Record* (19 June 1909), p. 792; "The Panama Canal - No. V," *Engineering* [London] (18 July 1913), pp. 70-71; and "Emergency Dam," *ASCE, Transactions*, p. 1632.

46   Curwood, *The Great Lakes* ..., pp. 28-64; and T.C. Keefer, "The Canals of Canada," pp. 40-44.

47   DRC, *Annual Report*, 1897, p. 119; and W.P. Kibbee, "The Busiest Canal in the World," *Engineering Magazine*, Vol. 13 (July 1897), pp. 600-610.

48   Curwood, *The Great Lakes* ..., pp. 28-64 and 121. Iron ore traffic through both 'Soo' canals doubled between 1900 and 1906, reaching 41 million tons

238 Technology in Transition

during the 1907 navigation season (ibid., pp. 28-29). Some 68 million bushels of wheat passed through in 1905, increasing to 84 million bushels in 1906 (ibid., p. 62). Coal transport also was growing rapidly with an anticipated eight million tons going to Duluth in 1909 (ibid., p. 64). Indeed, as of 1909 Duluth, the distribution point for coal going inland and grain going east, had surpassed London as the second busiest freight-shipping port in the world, second only to New York (ibid., p. 115).

49   During the 1890s, the largest of the upper lakes freighters were over 350 feet in length with less than a 43-foot beam (Keefer, "The Canals of Canada," p. 41). As of 1909, there were over 800 iron ore carriers on the upper lakes and the largest — "the giant of the Lakes" — was the 605 foot five inch long *Thomas F. Cole* with a 58-foot beam (Curwood, *The Great Lakes*, p. 19).

50   "The New United States Government Lock at Sault Ste. Marie, Mich." *Engineering News* (26 September 1895), pp. 194-196. The Canadian lock, however, remained the longest in the world until the August 1914 opening of the Panama Canal with its 1000 foot by 110 foot locks.

51   United States Army, *Annual Report of the War Department, Reports of Chief Engineers, U.S. Army*, Washington: Government Printing Office, "St. Mary's River at Falls, Michigan," 1897, Part IV, p. 2964 and 1898, Part IV, pp. 2554-2557, and "Work on the New Lock and Canal at Sault Sainte Marie, Mich.," *Engineering News, A Journal* (7 September 1911), p. 275.

When a freighter, the *Lakeshore*, struck the lock gates in the Poe lock in August 1910, the new swing bridge dam was under construction. Fortunately the upper gates, forced open by the freighter entering the lock from below, were driven closed again by the flow of water forcing the freighter back into the lock. This was described as a "miracle," which narrowly averted a disaster owing to their being no structure in place to cut off the flow of water out of Lake Superior had the lock gates been carried away, (Canal Records, File C-4256/S32-10, Vol. 2, newspaper clipping, "Miracle Averts Canal Disaster, Steamer *Lakeshore* Rams Poe Lock," 3 August 1910).

52   "Movable Dam at St. Mary's Falls Canal," *Engineering Record* (5 August 1911), p. 161. Although constructed in 1910-11, the American swing bridge dam had been designed and contracted out prior to the Canadian accident of June 1909 (see "The Accident at the Canadian Lock at Sault Ste. Marie," *Engineering Record* (19 June 1909), p. 792). Thus, the American improvements in the design of the Canadian prototype were not a direct result of the 1909 accident, unless design changes were made after the contract was let.

53   DRC, *Annual Report*, 1910, p. 264.

54   Ibid.; and Canal Records, File C-4256/S32-10, Vol. 2, J. Le B. Ross, Memorandum, 6 July 1909. There was no problem with the swing bridge su-

perstructure. During the whole operation in June 1909, the deflection of the centre of the lower downstream chord of the bridge had not exceeded 1-3/4 inch (Canal Records, ibid.).

55 Ibid., J. Le B. Ross to W.A. Bowden, Chief Engineer, DRC, 4 August 1910.

56 Ibid., George P. Graham, Minister, DRC, Memorandum, 9 February 1911.

57 Ibid., J. Le B. Ross to W.A. Bowden, 26 May 1911, and enclosed drawing, "Soo Canal Movable Dam, Repairs to Frame and Bracing," dated 20 December 1910.

58 Ibid., enclosed drawing, "Bevelled timbers inserted between wicket frames, 1911," December 1911.

59 J.D. Bouchard, a former canal superintendent, states that the steam hoist was installed following the 1909 accident (Bouchard to Norman Rutane, Chief Interpreter, Sault Ste. Marie Canal, 19 September 1985). The earliest reference found to the steam hoist is a 1919 inventory of canal structures stating that the wickets on the emergency dam were lowered by steam (Swainson and Osborne, *Sault Ste. Marie Canal...*, p. 139). As early as 1896, however, the canal superintendent had recommended that a portable steam boiler would be useful for operating the swing dam wicket hoist and other functions around the lock site (DRC, *Annual Report*, 1896, p. 98).

60 The steam hoist is described in: Canal Records, File C-4256/S32-10, Vol. 3, S. Hairsine, Sault Ste. Marie Canal, to J.B. McAndrew, Senior Assistant Engineer, Welland Canal, 11 January 1943; and ibid., C.W. West, Superintending Engineer, Welland Canal, to E.B. Jost, General Superintendent, Department of Transport, 13 August 1943.

61 "Power Transmission," *Electrical Engineer*, Vol. 20, No. 389 (16 October 1895), pp. 380-381; and E.G.M. Cape, "The Industries of the Consolidated Lake Superior Company," *Transactions of the Canadian Society of Civil Engineers*, Jan.-June 1903, pp. 154-155.

62 Canal Records, File 4256/S32-10, Vol. 4, R. L'Hereux, Structural Engineer, Memorandum, "Emergency Dam, Sault Ste. Marie," 7 January 1957.

63 McCullough, *The Path between the Seas, The Creation of the Panama Canal, 1870-1914*, pp. 482-488. Ferdinand de Lesseps had attempted, 1882-89, to excavate a sea level canal similar to his Suez Canal. The Americans, on commencing work in 1904, also had planned to excavate a sea level canal.

64 "The Panama Canal - No. V," *Engineering* [London] (18 July 1913), p. 70.

65 McCullough, *The Path between the Seas*, pp. 484 and 592.

66 Ibid., p. 539.

67 Goldmark, "Emergency Dam," *ASCE, Transactions*, p. 1632, Comments by T.B. Monniche, who designed the emergency swing bridge dams at Panama.

68   McCullough, *The Path between the Seas*, pp. 484-489, and 542. Hodges was assisted by Edward Schildnauer and Henry Goldmark (ibid., p. 594). The chief engineer on the Panama Canal, 1907 through 1914, was Lt. Col. George W. Goethals who also had experience in lock construction work. In 1889-94, he had built a lock of 26-foot lift — a then record lift — on the Cumberland and Shoals Canal (ibid., pp. 508-509).

69   Goldmark, "Emergency Dam," *ASCE, Transactions*, pp. 1635 and 1642, T.B. Monniche comments.

70   "The Panama Canal - No. V" *Engineering* (18 July 1913), p. 71.

71   Ibid., pp. 70-71; and McCullough, *The Path between the Seas*, pp. 599-600.

72   Goldmark, "Emergency Dam," *ASCE, Transactions*, p. 1592.

73   "The Panama Canal - No. V," *Engineering* [London] (18 July 1913), p. 72, and Figs. 113, 114, and 115.

74   Ibid.; Goldmark, "Emergency Dam," *ASCE, Transactions*, p. 1632; and "The Panama Canal - No. VI," *Engineering* [London] (1 August 1913), pp. 139-142. The wickets were constructed of nickel steel.

75   Ibid., p. 139; and Goldmark, "Emergency Dam," *ASCE, Transactions*, p. 1620.

76   Ibid., pp. 1618 and 1635; and "The Panama Canal - No. VI," *Engineering* [London] (1 August 1913), p. 141.

77   Goldmark, "Emergency Dam, *ASCE, Transactions*, p. 1619." A skeleton crew of three or four men could operate a Panama Canal swing bridge dam; but it required twice as much time to close the dam. With a large force of lock labourers on site, manpower was not really a critical concern on the Panama Canal (ibid., p. 1635).

78   Ibid., pp. 1622 and 1635.

79   Ibid., p. 1612.

80   Ibid., p. 1623.

81   "Work on the New Lock and Canal at Sault Sainte Marie, Mich.," *Engineering News, A Journal* (7 September 1911), pp. 275-279; "New Canal and Locks at 'The Soo'," *Engineering News, A Journal* (5 March 1914), pp. 512-519; and "New Lock and Canal at 'The Soo'," *Engineering News, A Journal* (23 April 1914), pp. 879-886.

82   "Work on New Lock and Canal," ibid., (7 September 1911), pp. 275-76.

83   L.C. Sabin, "New Type Movable Dam Guards Soo Canal Locks," *Engineering News-Record*, Vol. 93, No. 17 (1924), pp. 656-660.

84   Ibid., pp. 657-659.
         The designers were Isaac De Young, assistant engineer, and Owen M. Frederick, junior engineer, U.S. Army Corps of Engineers, Sault Ste. Marie.

The detailed bridge plans were prepared by the contractor, the Independent Bridge Company.

85   Ibid., pp. 657-660. Steam power was adopted to ensure operation during any emergency, independent of the functioning of the electrical power system used to operate the two locks.

86   W.J. Barden and A.W. Sargent, "The Lake Washington Ship Canal, Washington," *ASCE, Transactions,* Vol. 92, 1928, pp. 1001-108ff.; and Goldmark, "Emergency Dam," ibid., p. 1622.

87   Ibid., pp. 1622 and 1642.

88   "The Lake Washington Ship Canal," *ASCE, Transactions,* p. 1018.

89   Goldmark, "Emergency Dam," *ASCE, Transactions,* pp. 1592 and 1621. The idea of using stop-log girders originated with R.O. Comer, Designing Engineer, New Orleans Port Commission. Unbeknownest to Comer, a similar type of movable emergency stop-log dam had been constructed in Canada on the Trent Canal (ibid., p. 1592). The Trent Canal locks, however, were only 33 feet wide on a five-foot-deep barge navigation. Hence in designing emergency dams the Canadian engineers had faced nowhere near the immense scale of water control problems in contemplation at New Orleans.

90   Ibid., pp. 1592, 1643, and Fig. 25.

91   Ibid., pp. 1599-1602, and Fig. 7, "General Plan, New Orleans Emergency Dam."

92   Ibid., pp. 1608-1612, and Figs. 11 and 15.

93   Ibid., pp. 1601-1602.

94   Ibid., pp. 1612, 1619-1620, and 1643.

95   The consulting engineer for the New Orleans Inner Navigation Canal was George W. Goethals, the former chief engineer on the Panama Canal 1907-15. The plans for all structures on the New Orleans canal were prepared by Henry Goldmark, who had also worked on the design of the Panama Canal engineering structures (ibid., p. 1615).

96   Monniche, the designer of the Panama Canal swing bridge dams, continued to maintain that if stop-log dams were designed to meet the far greater demands made on the Panama Canal dams, they would not be any cheaper or more reliable in operation (ibid., pp. 1636-1641). In response, Goldmark admitted that the superiority of the stop-log dam might not hold for all widths and depths of locks (ibid., p. 1642).

97   The comparative merits of the two types of movable emergency dam are set forth in ibid., p. 1592, and the following discussion, pp. 1616-1645.

98   The movable emergency dam on the Lake Washington Ship Canal embodied the same wicket system and working principle for closing off the water as

the traditional swing bridge dams. The superstructure was, however, not a swing bridge.

99   See, "The Welland Ship Canal - XIII," *Engineering* [London] (18 July 1930), p. 63. The designer of the Welland Canal stop-log dam, F.E. Sterns, Designing Engineer, Welland Ship Canal, had previously worked as principal assistant engineer to Henry Goldmark in designing the New Orleans emergency stop-log dam (ibid.; and Goldmark, "Emergency Dam," *ASCE, Transactions*, p. 1615).

100   Personal Communication, Clarence C. Payne, Administrative Assistant, Panama Canal Commission, Balboa, Republic of Panama, to Robert W. Passfield, 7 January 1986, and enclosures: "Emergency Dams Will Be Offered for Sale," *The Panama Canal Review* (4 March 1955), p. 16; and excerpts from N.S. Chong, ed., *A History of the Panama Canal, From Construction Days to the Present* (Panama Canal Commission, 1984). Subsequently, the fender chains security system was also deemed superfluous. Most were removed in July 1976 and the remaining four, at the head of the Gatun and Pedro Miguel locks, in October 1980 (Chong, op. cit., p. 53).

101   The 800 foot by 80 foot MacArthur lock is protected by an emergency derrick/stop-log dam upstream of the upper guard gates. The dam consists of six 32-ton steel girders which are placed horizontally across the canal, in vertical grooves in the wall masonry, by a boom on an electrically powered stiff-leg derrick. So effective are stop-log girder emergency dams in tightly sealing off a canal that a second emergency stop-log girder dam was placed downstream of the lower guard gates. Both dams were built before the new lock chamber and used as coffer dams during its construction. (See J.R. Carr, "MacArthur Lock at the Soo," *Engineering News-Record* (18 November 1943), pp. 78-85; and "The MacArthur Lock at Sault Ste. Marie," *Engineering* [London] (28 April 1944), pp. 321-323.

102   DRC, *Annual Report*, 1920-21, p. 85; ibid., 1929-30, p. 105; and ibid., 1939-40, p. 49. The swing bridge dam was painted in all three of these years. On the end rest girder see, Canal Records, File C-4256/S32-10, Vol. 3, blueprint, Department of Transport, Sault Ste. Marie Canal, "New End Rest Girder for Steel Swing Dam," 16 January 1939.

      The subtitle quote is from Goldmark, "Emergency Dam," *ASCE, Transactions*, p. 1616.

103   Canal Records, File C-4256/S32-10, Vol. 4, G.N. Phillips, Superintending Engineer, Sault Ste. Marie Canal, to R.J. Burnside, Director, Canal Services, Department of Transport, 15 February 1956. In 1936, the Department of Transport superseded the Department of Railways and Canals.

104 Ibid., Vol. 3, J.B. McAndrew, Assistant Engineer (Hydraulic), St. Catharines, Ontario, "Report to E.B. Jost on the Condition of Mechanical and Structural Equipment of the Canadian Lock at Sault Ste. Marie," 24 November 1942; and ibid., J.C. McLeod, Superintending Engineer, Sault Ste. Marie Canal, to E.B. Jost, General Superintendent of Canals, DOT, 14 January 1943.

105 Osborne and Swainson, *The Sault Ste. Marie Canal...*, pp. 121-123.

106 Canal Records, File C-4256/S32-10, Vol. 3, McAndrew to Jost, 24 November 1942.

107 Ibid., C.W. West, Superintending Engineer, Welland Canal, to E.B. Jost, General Superintendent of Canals, DOT, 11 June 1943.

108 Ibid., Vol. 1, Robert C. Douglas, Hydraulic and Bridge Engineer, DRC, to C. Schreiber, 30 April 1896. The original specifications had called for the bridge to be swung by four men (ibid., Sault Ste. Marie Canal, "Specification, For a Movable Dam and Swing Bridge of Steel," 15 January 1895). The rim-bearing swing bridge designed by Dominion Bridge, however, was exceptionally well balanced and easy to turn. The much smaller American swing bridge dam, the 1885 structure, had required six men to swing with its hand-operated turning gear ("St. Mary's Falls Canal Bridge and Dam," *Engineering Record* [22 March 1890], p. 246).

109 Canal Records, Vol. 3, File C-4256/S32-10, McAndrew to Jost, 24 November 1942. J.C. Macleod, Superintending Engineer, Sault Ste. Marie Canal, admitted that the swing bridge turntable had not been greased for years and agreed that the swing bridge dam, if electrically powered, could be fully operated at least once each year during the navigation season without seriously delaying shipping (ibid., Macleod to Jost, 14 January 1943).

110 Ibid., West to Jost, 9 August 1943; and ibid., West to Jost, 13 August 1943.

111 Ibid., West to Jost, 13 August 1943.

112 Ibid., Provincial Engineering Ltd., Niagara Falls, Ontario, was prepared to provide an electrically powered chain block of the type required for $1440 (ibid.).

113 Ibid., West to Jost, 11 June 1943 and 9 August 1943. On the basis of eight men being needed to start the swing bridge in motion, and in assuming a load of 125 pounds at a five-foot leverage on the turning lever, McAndrew calculated that the old cast iron turning gears and shafts were stressed at about 80 percent of their breaking stress. Hence, as a safety precaution, he recommended that new cut steel gears be installed (ibid., West to Jost, 11 June 1943).

114 Ibid., D.W. McLachlan, Memorandum to Jost, 3 September 1943.

115 Ibid., West to Jost, 9 August 1943.

116 A.G. Ballert, "The Soo Versus the Suez," *Canadian Geographical Journal*, Vol. 53 (November 1956), pp. 160-167. In the post-war period, the American canals carried 97 percent of the total freight tonnage through the 'Soo.' The Canadian canal, however, dominated the passenger vessel trade, and retained a significant role in the grain traffic (ibid.).

117 Canal Records, File C-4256/S32-10, Vol. 3, H.M. Campbell for Superintending Engineer, Sault Ste. Marie Canal, to J.H. Ramsay, Acting Director, Canal Services, DOT, 8 February 1947.

118 Ibid., G.N. Phillips, Superintending Engineer, to C.W. West, Director, Canal Services, 12 July 1950. The underwater sill was also cleared of debris and inspected by a diver during the July 1950 inspection (ibid.).

119 Ibid., C.W. West, Memorandum to S. Hairsine, 28 August 1950.

120 Ibid., E.C. Shurly, Report on Emergency Dam, 8 September 1950.

121 Ibid., Report on Emergency Dam, 8 September 1950; and ibid., L.H. Burket, Chief Assistant, Structural Sales, Dominion Bridge Company, to C.W. West, Director, Canal Services, DOT, 19 September 1950.

122 Ibid., C.W. West, Memorandum, "Transfer Between Allotments," to Director, Administration and Legal Services, 6 October 1950. On the Rideau Canal and the Trent Canal, monies had been saved through reducing the operating hours from 24 to 12 hours-per-day, and revising the Sunday hours. Over the course of the 1950 navigation, $18 250 had been saved on the Rideau, and $30 000 on the Trent, through a corresponding decrease in the respective salaries of the permanent canal staffs (ibid.).

123 Ibid., Department of Transport, Sault Ste. Marie Canal, Specification, 14 December 1950.

124 Ibid., G.N. Phillips, Superintending Engineer, Sault Ste. Marie Canal, to C.W. West, Director, Canal Services, 1 February 1951; and ibid., E.C. Shurly, for Director, Canal Services, to G.N. Phillips, 5 February 1951.

125 Ibid., D.B. Armstrong, Assistant Chief Engineer, Dominion Bridge Company, to C.W. West, 16 February 1951; and ibid., West to Armstrong, 21 February 1951. Dominion Bridge recommended the use of bronze bearings and, in response, DOT specified the use of Oilite bearings produced by the Acme Machine and Tools Ltd. in Toronto.

126 Ibid., West to E.A. Kelly, Sault Structural Steel Co. Ltd., 6 March 1951; and Ibid., D.B. Armstrong, Dominion Bridge, to West, 9 March 1951. In the annealing or "normalizing" process, the chains were heated beyond a critical point and then cooled in air to soften the steel links and make them less brittle.

127 Ibid., G.N. Phillips, Superintending Engineer, Sault Ste. Marie Canal, to C.W. West, 1 February 1951; and ibid., West to Kelly, Sault Structural Steel, 6 March 1951.

128 Ibid., C.W. West, Director, Canal Services, Memorandum to File, 9 February 1951; and E.A. Kelly, Sault Ste. Marie Structural Steel Company, to West, 12 February 1951.

129 Ibid., West to Kelly, 6 March 1951; and G.N. Phillips, Superintending Engineer, Sault Ste. Marie, Telegram to West, 26 April 1951.

130 Ibid., D.B. Armstrong, Dominion Bridge, to C.W. West, 16 February 1951.

131 Ibid., Armstrong to West, 23 February 1951.

132 Ibid., J.E. Garrigan, Chief Draughtsman, Dominion Bridge, to C.W. West, 13 March 1951; and ibid., West to Garrigan, 16 March 1951.

133 Ibid., G.N. Phillips, Telegram to C.W. West, 26 April 1951.

134 Ibid., C.W. West to E.A. Kelly, Sault Structural Steel, 6 March 1951.

135 Ibid., West to Phillips, 30 April 1951.

136 Ibid., Vol. 4, F.T. Shearns, for Director, Cost Inspection and Audit Division, 14 January 1952; and ibid., E.M. Clough, Cost Auditor, Treasury Cost Auditor's Supplementary Report, 18 March 1952. Dominion Bridge had requested payment of $55 179.84 to cover costs plus a 15 percent profit and federal taxes. (ibid., Dominion Bridge to DOT, 20 July 1951).

137 Ibid., Vol. 3, Transfer between Allotments, DOT, 22 March 1951; and Ibid., E.C. Shurly, for Director Canal Services, to G.N. Phillips, Superintending Engineer, Sault Ste. Marie Canal, 14 June 1951. A sum of $2 982 350 had been allotted to construct a tunnel under the Lachine Canal at St. Remi. All of the allotted funds, however, had not been used up by the progress of the work at the end of the fiscal year on 31 March 1951.

138 Ibid., W.F. Walker, General Manager, Provincial Engineering, to J.B. McAndrew, Welland Ship Canal, St. Catharines, 29 August 1950. The hoist operated on a three-phase, 60-cycle power supply. McAndrew was consulted because of his earlier electrification plan, see S. Hairsine, for Director, Canal Services, to J.H. Ramsay, Superintending Engineer, Welland Canals, 25 August 1950.

139 Ibid., Vol. 4, J.G. Taylor, Technician, Memorandum to G.N. Phillips, Superintending Engineer, 17 July 1951.

140 Ibid., E.A. Kelly, Manager, Sault Structural Steel, to G.M. Phillips, 2 September 1952.

141 Phillips to C.W. West, 10 August 1953.

142 Ibid., Vol. 4, R. L'Heureux, Structural Engineer, Emergency Dam, Sault Ste. Marie Canal, 7 January 1957. The steam hoist crew consisted of: the hoist operator, a fireman for the boiler, a helper for handling the cables at the hoist,

a signalman, a man handling the chain stop, and five labourers to handle the wicket chain and move the snatch block pulleys. The electric hoist required only an electrician, and three helpers to handle the wicket chain and chain stop.

143 Personal Communication, Norman Rutane, Chief of Interpretation, Sault Ste. Marie Canal, to R.W. Passfield, 19 September 1985. The old boiler remains on the swing bridge dam. The new portable boiler was used also for powering the gate lifter when stepping or unstepping lock gates (ibid.).

144 Canal Records, Headquarters, Acquisitions, Sault Ste. Marie Canal, General, File C-8616/S32, Vol. 1, order-in-council, P.C. 1959-204, Transfer, effective 1 April 1959.

145 Ibid., Vol. 9, J.D. Bouchard, Superintending Engineer, Sault St. Marie Canal, "Emergency Dam," 4 July 1978; and Sault Ste. Marie Canal, Canal Office, Engineering Drawings, Case X, Drawing No. 50421, "Replace Floor Beams, Stringers, and Counterweight on Emergency Dam," 10 January 1971. The building housing the old boiler on the swing bridge deck was also renewed at this time.

146 Canal Records, Headquarters, File C-8616/S32, Vol. 1, John I. Nicol, Director-General, to A.T. Davidson, Assistant Deputy Minister, 2 March 1978; and ibid., File C-8500/S32, Realty, Sault Ste. Marie Canal, Vol. 1, Task Force Report, *Sault Ste. Marie Lock and Adjacent Federal Lands, Preliminary Assessment of Options*, October 1977, pp. 2, 5-6.

147 Environment Canada, Canadian Parks Service, Ontario Region, *Sault Ste. Marie Heritage Canal, Interim Management Plan*, 1982, pp. 1-2.

148 Canal Records, File C-4272-S32, Vol. 1, T.J. Kearney, Marine Work and Transportation, Engineering and Architecture Branch, Parks Canada, to D.B. Bazely, Manager, Fenco Engineering Ltd., Sault Ste. Marie, 7 March 1983.

REFERENCES CITED

American Architect and Building News
"Great Canals of the World." Vol. 75 (January-March 1902), p. 47. New York.

Atkinson, Philip
The Elements of Electric Lighting, including Electric Generation, Measurement, Storage and Distribution. D. Van Nostrand Company, New York, 1890.

Ballert, Albert G.
"The Soo versus the Suez." *Canadian Geographical Journal*, Vol. 53 (November 1956), pp. 160-167.
——. "Commerce of the Sault Canals." *Economic Geography*, Vol. 33 (April 1957), pp. 135-148.

Barden, W.J., and A.W. Sargent
"The Lake Washington Ship Canal, Washington." *American Society of Civil Engineers Transactions (ASCE Transactions)*, Vol. 92, Paper No. 1679 (1928), pp. 1001-1018.

Barry, James P.
Ships of the Great Lakes, 300 Years of Navigation. Howell-North Books, Berkeley, California, 1974.

Buckley, W.J.
Electric Lighting Plants, their Cost and Operation. William Johnston Printing Co., Chicago, 1894.

Bush, Edward F.
"A History of Hydro-Electric Development in Canada." Microfiche Report Series, No. 306, Environment Canada, Canadian Parks Service, Ottawa, 1987.

Canada. Department of Railways and Canals (DRC)
Annual Report. Queen's/King's Printer, Ottawa, 1899-1936.

Canada. Department of Transport
Annual Report. King's Printer, Ottawa, 1936-1956.

Canada Lumberman
"The Electric Light, Mr. E.B. Eddy's New Enterprise." (1 July 1881), p. 5. Toronto.

Canada. Parliament. House of Commons

House of Commons Debates. Queen's Printer, Ottawa, 1888, Vol. 2, pp. 1443-1444.

———. *Journals of House of Commons, 1895,* Vol. 29, Appendix, "Sault Ste. Marie Canal Inquiry, Minutes of Evidence," Queen's Printer, Ottawa, 1896.

———. *Report of the Auditor General.* King's Printer, Ottawa, 1906. Vol. 3, p. W-118, "Sault Ste. Marie Canal: Repairs," and 1921, Vol. 3, p. W-108, "Sault Ste. Marie Canal: Repairs."

Canadian Engineer (Toronto)

"The Sault Ste. Marie Ship Canal." Vol. 1 (November 1893), pp. 191-192.

———. John D. Barnett, "Pneumatic Power in Workshops." Vol. 4 (July 1896), pp. 61-64.

———. "The Taylor Hydraulic Air Compressor." Vol. 2 (April 1895), pp. 343-346.

———. "The Taylor System of Air Compression." Vol. 4 (November 1896), p. 194.

———. "Hydraulic Air Compressor at Magog." Vol. 4 (March 1897), p. 316.

———. "Canadian Reinforced Concrete Arch Bridges." (13 March 1919), pp. 289-293.

———. "Operation of Locks by Electrical Power — The New Experiments on the Beauharnois Canal." Vol. 1 (August 1893), pp. 91-92.

———. "The Montreal Street Railway Power House." Vol. 1 (January 1894), p. 252.

———. "Electrical Development." Vol. 1 (January 1894), p. 249.

———. John Langton, "Direct Connected Dynamos with Steam Engines." Vol. 1 (October 1893), pp. 163-164.

———. A.C. McCallum, "Turbine Water Wheels." Vol. 1, No. 6 (1893), pp. 146-148.

———. "Montmorency Falls Electric Plant." Vol. 4 (June 1896), pp. 50-53.

———. "The Multiphase System of Electricity." Vol. 1 (July 1893), pp. 65-66.

———. F.C. Armstrong, "Three-Phase Transmission." Vol. 4 (March 1897), pp. 322-323.

———. "Water Power at Niagara Falls." Vol. 2 (November 1894), pp. 198-200.

———. J.C. Sing, "Canada and Her Waterways." (22 January 1909), pp. 142-144.

———. "The Movable Dam and Swing Bridge on the Sault Ste. Marie Canal." Vol. 4, No. 4 (August 1896), p. 98.

Canadian Journal of Fabrics

"A Canadian Cotton Mill." (December 1897), pp. 367-370. Montreal.

Canal Records

Environment Canada, Canadian Parks Service, Ontario Region, Cornwall, Ontario. File C-4250/S32-1. "Sault Ste. Marie Canal, Construction, Maintenance and Repair, General." Vols. 1-11.

———. File C-4272-S32. "Sault Ste. Marie Canal, Construction, Maintenance and Repair, Lock and Gates." Vols. 1-8.

———. File C-4256/S32-10. "Sault Ste. Marie Canal, Dams and Weirs, Swing Dam." Vols. 1-10.

———. Central Registry, Ottawa, Ontario. File C-8616/S32. "Acquisitions, Sault Ste. Marie Canal, General." Vol. 1.

Cape, E.G.M.
"The Industries of the Consolidated Lake Superior Company." *Transactions of the Canadian Society of Civil Engineers*, Vol. 17, Paper No. 177 (January-June 1903), pp. 150-170.

Carr, J. Roland
"MacArthur Lock at the Soo." *Engineering News-Record: A Journal of Civil Engineering and Construction* (18 November 1943), pp. 78-85. New York.

Carter-Edwards, Dennis
"The Sault Ste. Marie Canal." *Research Bulletin*, No. 119, Parks Canada, Ottawa, 1980.

Chong, N.S., ed.
A History of the Panama Canal, From Construction Days to the Present. Library Services Branch, Panama Canal Commission, Balboa Heights, Republic of Panama, 1984.

Coerper, C.
"The Electric Lighting of the North Sea and Baltic Canal." *The Electrical Review*, Vol. 36, No. 917 (21 June 1895), pp. 763-767. London.

Condit, Carl W.
American Building, Materials and Techniques from the Colonial Settlements to the Present. University of Chicago Press, Chicago, 1968.

———. "The Pioneer Stage of Railroad Electrification." *Transactions of the American Philosophical Society*, Vol. 67, Part 7, 1977, pp. 3-45.

Coutts, Sally
"Sault Ste. Marie Canal Buildings, Sault Ste. Marie, Ontario." Building Report: 85-07, Federal Heritage Buildings Review Office, Environment Canada, Canadian Parks Service, 1985.

Crocker, Francis B.
"The History of Electric Lighting." *The Electrical World*, Vol. 27 (9 May 1896), pp. 511-513. New York.

Curwood, James O.
The Great Lakes, The Vessels that Plough Them: Their Owners, Their Ships, Their Sailors, and Their Cargoes. G.P. Putnam & Sons, New York, 1909.

Cuthbertson, George A.
Freshwater: A History and a Narrative of the Great Lakes. MacMillan, Toronto, 1931.

Dales, John H.
Hydroelectricity and Industrial Development, Quebec 1898-1940. Harvard University Press, Cambridge, Massachusetts, 1957.

Davey, Norman
A History of Building Materials. Phoenix House, London, England, 1961.

Davis, Allan R.
"The St. Lawrence Canal Route." Canadian Magazine, Vol. 3 (1894), pp. 148-153.

Denis, Leo G.
Electric Generation and Distribution in Canada. Commission of Conservation, Ottawa, 1918.

Duncan, Louis
"Present Status of the Transmission and Distribution of Electrical Energy." *Annual Report of the Board of Regents of the Smithsonian Institution*, Government Printing Office, Washington, 1896, pp. 207-221.

Easterbrook, W.T., and H.G.J. Aitken
Canadian Economic History. Macmillan, Toronto, 1958.

Electrical Engineer, A Weekly Review of Theoretical and Applied Electricity
"Power Transmission, The Canadian Ship Canal Lock at Sault Ste. Marie and Its Electrical Operation." Vol. 20, No. 389 (16 October 1895), pp. 380-381. London.

Electrical World
"Electrical Equipment of Charles River Locks, Extensive electric motor application at the lower end of Charles River Basin for operating boat locks connecting the basin with Boston Harbor." Vol. 61, No. 25 (21 June 1913), pp. 1355-1361. New York.

Encyclopedia Britannica
"Lighting - Arc Lamps." 11th Edition (1910-11), pp. 665-666.

Engineering, An Illustrated Weekly Journal **(London)**
"The Manchester Ship Canal." Part I (25 April 1890), p. 497, Part II (30 May 1890), pp. 640-641, and Part III (6 June 1890), pp. 682-684.
———. "The Panama Canal - No. V." (18 July 1913), pp. 70-72. London.
———. "The Panama Canal - No. VI." (1 August 1913), pp. 136-142.
———. "The Welland Ship Canal - XIII." (18 July 1930), pp. 63-65.
———. "The MacArthur Lock at Sault Ste. Marie." Vol. 157 (28 April 1944), pp. 321-323.

Engineering and Building Record and Sanitary Engineer **(New York)**
"St. Mary's Falls Canal Bridge and Dam." Vol. 22 (22 March 1890), pp. 246-248.
———. "Accident at the Canadian Lock at Sault Ste. Marie." Vol. 59, No. 25 (19 June 1909), pp. 789-792.
———. "Movable Dam at St. Mary's Falls Canal." Vol. 64 (5 August 1911), p. 161.
———. The North-Baltic Canal Locks, Holtenau." Vol. 31, No. 22 (27 April 1895), pp. 381-382.
———. "The Canadian Ship Canal at Sault Ste. Marie," Vol. 32, No. 25 (16 November 1895), pp. 435-436.

Engineering and Mining Journal
"Accident to the Sault Ste. Marie Canal Lock." Vol. 50 (30 August 1890), pp. 241-242. New York.

Engineering News and American Railway Journal **(New York)**
"The Canadian Ship Canal Lock at Sault Ste. Marie, Ont." Vol. 33, No. 13 (28 March 1895), pp. 205-207.
———. "The Canadian Lock at Sault Ste. Marie." Vol. 33, No. 25 (20 June 1895), pp. 398-399.
———. "Steel Lock Gates for 800 x 100 Ft. Ship Canal Lock, Sault Ste. Marie, Mich." Vol. 36, No. 6 (6 August 1896), pp. 84-86.
———. "The Illinois & Mississippi Canal Lock Works." Vol. 33, No. 7 (14 February 1895), pp. 98-101.
———. "The Merits of Rubble Concrete." Vol. 50, No. 3 (16 July 1903), pp. 58-59.
———. The North Sea and Baltic Ship Canal." Vol. 33, No. 25 (20 June 1895), p. 398.
———. "The New United States Government Lock at Sault Ste. Marie, Mich." Part I, Vol. 34, No. 13 (26 September 1895), pp. 194-196.
———. "The New United States Government Lock at Sault Ste. Marie, Mich." Part II, Vol. 34, No. 15 (10 October 1895), pp. 238-239.

——. E.S. Wheeler, "Locks of the Nicaragua Canal and St. Mary's Falls Canal." (1 June 1893), pp. 504-505.

——. C.R. Coutlee, "The Soulanges Canal Works, Canada." Part I, Vol. 45, No. 16 (18 April 1901), pp. 274-278.

——. C.R. Coutlee, "The Soulanges Canal Works, Canada." Part II, Vol. 46, No. 2 (11 July 1901), pp. 30-32.

Engineering News, A Journal of Civil, Mechanical, Mining and Electrical Engineering **(New York)**
"The Accident to the Lock Gates at the Canadian Canal, Sault Ste. Marie." Vol. 61, No. 24 (17 June 1909), pp. 672-674.

——. "Work on the New Lock and Canal at Sault Sainte Marie, Mich." Vol. 66, No. 10 (7 September 1911), pp. 275-279.

——. "New Canal and Locks at 'The Soo'. " Vol. 71, No. 10 (5 March 1914), pp. 512-519.

——. "New Lock and Canal at 'The Soo'. " Vol. 71, No. 17 (23 April 1914), pp. 879-886.

Friesen Gerald
The Canadian Prairies, A History. University of Toronto Press, Toronto, 1984.

Frizell, Joseph P.
Water-Power, An Outline of the Development and Application of the Energy of Flowing Water. 2nd ed. John Wiley & Sons, New York, 1901.

Fowke, Vernon C.
The National Policy and the Wheat Economy. University of Toronto Press, Toronto, 1957.

Gillespie, Ann
"Early Development of the Artistic Concrete Block: The Case of the Boyd Brothers." *APT Bulletin*, Vol. 11, No. 2 (1979), pp. 30-52.

Globe **(Toronto)**
"The Ship Canal." 13 November 1890.

——. "The Canadian Soo and the Great Canal." 26 October 1895, p. 2.

Globe and Mail **(Toronto)**
"New Lakes Era Marked by Freighter." 4 May 1972.

Goff, John H.
"History of the St. Mary's Falls Canal." *The Saint Marys Falls Canal.* Charles Moore, ed. 1905 Semi-Centennial Commission, Detroit, 1907. pp. 127-168.

Goldmark, Henry
"Emergency Dam on Inner Navigations Canal at New Orleans, Louisiana." *American Society of Civil Engineers Transactions (ASCE Transactions)*, Vol. 92, Paper No. 1695 (1928), pp. 1589-1645. New York.

Hammond, John W.
Men and Volts, The Story of General Electric. T.B. Lippencott Co., New York, 1941.

Hatcher, Harlam, and Erich A. Walter
A Pictorial History of the Great Lakes. Crown Publishers, New York, 1963.

Heisler, John P.
The Canals of Canada. Parks Canada, Ottawa, 1973.

Herdt, L.A.
"The Use of Electricity on the Lachine Canal." *Transactions of the Canadian Society of Civil Engineers (Transactions, CSCE)*, Paper No. 207 (24 March 1904), pp. 161-172. Montreal.

Hughes, Thomas P.
Network of Power, Electrification in Western Society 1880-1930. John Hopkins University Press, Baltimore, Maryland, 1983.

Hunter, Louis C.
A History of Industrial Power in the United States, 1789-1930, Volume One: Waterpower in the Century of the Steam Engine. University Press of Virginia/Eleutherian Mills-Hagley Foundation, Charlottesville, *Virginia*, 1979.

Innis, Harold A.
A History of the Canadian Pacific Railway. University of Toronto Press, Toronto, 1923 (reprinted 1971).

Johnston, Archibald F.
Canadian General Electric's First Hundred Years, A Chronological Sketch. Canadian General Electric Co., Toronto, 1982.

Keefer, Thomas C.
"The Canals of Canada." *Transactions of the Royal Society of Canada*, 1893, Section III, pp. 25-50.

Kibbee, William P.
"The Busiest Canal in the World." *Engineering Magazine*, Vol. 13 (July 1897), pp. 600-614. New York.

Kuttruff, Karl, R.E. Lee, and D.T. Glick
Ships of the Great Lakes: A Pictorial History. Wayne State University, Detroit, 1976.

Lawson, A.J.
"Generation, Distribution and Measurement of Electricity for Light and Power." *Transactions, Canadian Society of Civil Engineers (Transactions CSCE)*, Vol. 4, Paper No. 43 (1890), pp. 179-240.

Leonard, F.H.
"Electrical Equipment for Cornwall Canal." *Transactions of the Canadian Society of Civil Engineers*, Paper No. 208 (24 March 1904), pp. 173-187.

Leung, Felicity L.
"Direct Drive Waterpower in Canada: 1607-1910." Microfiche Report Series, No. 271, Environment Canada, Canadian Parks Service, Ottawa, 1986.

MacGibbon, D.A.
The Canadian Grain Trade. Macmillan, Toronto, 1932.

Marine Review
"Third Lock at Sault." (December 1914), pp. 468-473. Cleveland, Ohio.

Martin, Thomas C., and Joseph Wetzler
The Electric Motor and Its Application. 2nd ed. W.J. Johnston Publisher, New York, 1888.

McCullough, David G.
The Path between the Seas, The Creation of the Panama Canal, 1870-1914. Simon & Shuster, New York, 1977.

Meyer, Herbert W.
A History of Electricity and Magnetism. MIT Press, Cambridge, Massachusetts, 1971.

Monniche, T.B.
"Emergency Dams Above Locks of the Panama Canal." *Transactions of the International Engineering Congress, 1915*, Vol. 2 (1916), pp. 131-164. San Francisco.

Moore, Charles, ed.
The Saint Mary's Falls Canal. 1905 Semi-Centennial Commission, Detroit, 1907.

Morton, Desmond
A Short History of Canada. Hurtig Publishers, Edmonton, 1983.

Murphy, M.
"Concrete as a substitute for Masonry in Bridge Work." *Transactions, Canadian Society of Civil Engineers (Transactions, CSCE),* Vol. 2, Paper No. 14 (23 February 1888), pp. 79-111.

Nanne, Glorya
"Canals: Concrete and Steel Replace Limestone in Soo Lock Walls." *Public Works in Canada* (22 May 1962), pp. 28-29.

National Archives of Canada (NA)
RG43, Records of the Department of Railways and Canals, Vols. 1697-1698.
——. RG43, B2e, Department of Railways and Canals, Canal Records, Office of the Chief Engineer, Canal Branch, Contract Records, Vols. 1771-1772.
——. RG43, B61, Department of Railways and Canals, Rideau Canal, Office of the Superintending Engineer, Letterbook, Vols. 2009-2013.

Nelson, Daniel
Managers and Workers, Origins of the New Factory System in the United States, 1880-1920. University of Wisconsin Press, Madison, 1975.

Osborne, Brian S., and Donald Swainson
The Sault Ste. Marie Canal, A Chapter in the History of Great Lakes Transport. Environment Canada, Canadian Parks Service, Ottawa, 1986.

Owram, Doug
Promise of Eden: The Canadian Expansionist Movement and the Idea of the West, 1856-1900. University of Toronto Press, Toronto, 1981.

Panama Canal Review
"Emergency Dams Will Be Offered for Sale." (4 March 1955), p. 16. Panama.

Passer, Harold C.
The Electrical Manufacturers, 1875-1900, A Study in Competition, Entrepreneurship, Technical Change, and Economic Growth. Harvard University Press, Cambridge, 1953 (reprinted 1978).

Passfield, Robert W.
Building the Rideau Canal: A Pictorial History. Fitzhenry & Whiteside, Toronto, 1982.
——. "Canal Lock Design and Construction: The Rideau Experience, 1826-1982." Microfiche Report Series, No. 57, Environment Canada, Canadian Parks Service, Ottawa, 1983.

Porter, Arthur
"Electric Power." *The Canadian Encyclopedia*, Vol. 1 (1985), pp. 555-557.

Ritchie, Thomas
Canada Builds, 1867-1967. University of Toronto Press, Toronto, 1967.

Roos, Arnold E.
"Working Paper on Hydro-Electric Technology." Historical Research Branch, Environment Canada, Canadian Parks Service, Ottawa, 1985.

Ross, J.W. Le B.
"General Design of a Lock and Approaches." *Journal of the Engineering Institute of Canada*, Vol. 3 (August 1920), pp. 383-386. Montreal.

Ross, W.G.
"Development of Street Railways in Canada." *Canadian Magazine*, Vol. 18 (January 1902), pp. 276-278. Toronto.

Sabin, L.C.
"New Type Movable Dam Guards Soo Canal Locks." *Engineering News-Record: A Journal of Civil Engineering and Construction*, Vol. 93, No. 17 (1924), pp. 656-660.

Sault Daily Star **(Sault Ste. Marie, Ont.)**
"May Re-open Today, Lock Nearly Ready for Business Again." 17 June 1909.
——. "Traffic at Sault Canal Shows Heavy Increase over 1912." 13 September 1913.

Sault Ste. Marie Ship Canal Office
Environment Canada, Canadian Parks Service, Sault Ste. Marie, Ontario, Historic Photographs Collection.
——. *Ross Notebook*, n.d.

Scientific American Supplement **(New York)**
"The Sault Ste. Marie Canal." Vol. 30, No. 766 (6 September 1890), p. 12231.

Sellon, R. Percy
"Electric Light Applied to Night Navigation upon the Suez Canal." *The Telegraphic Journal and Electrical Review* (14 September 1888), pp. 279-282. London.

Stanley, Christopher C.
Highlights in the History of Concrete. Cement & Concrete Association, London, England, 1979.

Strandh, Sigvard
A History of the Machine. A. & W. Publishers, New York, 1979.

Thompson, Elihu
"Electrical Advance in the Past Ten Years." *Annual Report of the Board of Regents of the Smithsonian Institution,* Government Printing Office, Washington, 1897, pp. 125-136.

The Times Atlas of World History
Hammond Inc., Maplewood, New Jersey, 1978, pp. 222-223.

Tomlin, Edwin
"The Place of Concrete in the Coming Era, On the North American Continent." *Concrete & Constructional Engineering* (1920), pp. 20-22 and 101-102.

Tyler, W.W.
"The Evolution of the American Type of Water Wheel." *Journal of the Western Society of Engineers,* Vol. 3, No. 2 (April 1898), pp. 879-901. Chicago.

United States Army. Corps of Engineers
Annual Report of the Chief Engineers to the Secretary of War for the Year 1885. Part III, Government Printing Office, Washington, 1886, p. 2111.

Upp, John W.
"The Electric Operation of the Panama Canal Locks." *Sibley Journal of Engineering,* Vol. 28, No. 6 (March 1914), pp. 211-223. Ithica, N.Y.

Van Every, Margaret
"Francis Hector Clergue and the Rise of Sault Ste. Marie as an Industrial Centre." *Ontario History* (September 1964), pp. 191-202. Toronto.

Waite, Peter B.
Canada 1874-1896, Arduous Destiny. McClelland and Stewart, Toronto, 1971.

Walter, E.A.
A Pictorial History of the Great Lakes. Crown Publishers, New York, 1963.

White, Hon. Peter
"Historical Address." *The St. Mary's Falls Canal.* Charles Moore, ed. 1905 Semi-Centennial Commission, Detroit, 1907. pp. 31-32.

Yeater, Mary
"The Hennepin Canal." *American Canals, Bulletin of the American Canal Society.* No. 22 (August 1977), p. 7.

Young, C.R.
"Bridge Building." *The Engineering Journal*, Vol. 20, No. 6 (June 1937), pp. 478-499.

INDEX

61795

Passfield, Robert W

Technology in transition:
the 'Soo' ship canal, 1889-
1985.